The Mason's Toolbox Manual

David Tenenbaum

ARCO

New York London Toronto
Sydney Tokyo Singapore

is designed to provide accurate and authoritative information in
object matter covered. It is sold with the understanding that the
author do not render engineering or architectural advice and no
expressed or implied, are made as to the material contained herein. If expert
s required, contact a competent professional person.

 ARCO

Simon & Schuster, Inc.
15 Columbus Circle
New York, NY 10023

DISTRIBUTED BY PRENTICE HALL TRADE SALES

Manufactured in the United States of America

1   2   3   4   5   6   7   8   9   10

Library of Congress Cataloging–in–Publication Data

Tenenbaum, David.
    Mason's toolbox manual / David Tenenbaum.
        p.   cm.
    ISBN 0-13-558875-8
    1. Masonry.   I. Title.
    TH5313.T46   1990   90-33587
    693^{1}.1—dc20   CIP

Part One
Your Tools and Materials

1 Hand Tools / 1

Brushes / 1
Chalk Line / 3
Chisel / 3
Cutter / 4
Float / 4
Hammer / 5
Jointer / 5
Layout Tools / 6
  *Builder's level, Transit level,
  Laser transit, Selecting a level
  or transit*
Level / 10
Line and Line Stretcher / 11
Miscellaneous Tools / 12
Masonry Guide (Corner
  Pole) / 13
Mixing Equipment-Hand / 14
Mortar Board or Tub / 14
Plumb Bob / 15
Rulers / 15
  Brick spacing and modular
Safety Equipment / 16
Square / 17
Stone Tools / 17
Story Pole / 18
Tongs, Brick and Block / 18
Trowel / 18

2 Power Tools / 22

Electric Cord / 22
Drill / 22
Forklift / 23
Grinder / 23
Hoist / 24
Mortar Mixer / 24
Power Washer / 24
Sandblaster / 24
Saw / 25

Stud Tool / 25
Wheelbarrow and Carts / 26

3 Scaffolding and Ladders / 27

General Scaffold Safety / 27
Botswain's Chair / 28
Chimney Scaffold / 28
Horses and Bucks / 29
Ladder / 29
Rolling Tower / 30
Morgan Tower / 31
Swing Stage / 31
Tubular / 32

4 Materials / 33

Anchors / 33
Block, Concrete / 38
Block, Chimney / 45
Block, Glass / 45
Brick / 46
  *Absorbency, Grade, Size,
  Terminology, Types*
Caulking / 54
Coatings / 56
  *Bentonite, Bituminous sealers,
  Elastic compounds, Expansion
  compounds, Hydraulic cement,
  Silicon sealer, Choosing a coating,
  Paint*
Concrete / 60
Coping / 61
Flashing / 62
Flue Liner / 63
Insulation / 64
Movement Joint / 66
  Control Joint, Expansion Joint
Mortar / 68

Desired properties, Ingredients,
Special-purpose cements and
mortars
Paving Material / 79
Portland Cement / 80
Reinforcing and Ties / 81
   Angle iron, I-beam, Reinforcing

bar, Steel plate, Strap anchor, Joint
Reinforcing
Stone Masonry / 85
Terra Cotta / 87
Tile / 88
Wall Ties / 90

Part Two
Walls

5 Wall Types / 94

Cavity / 94
Composite / 101
Foundation and Basement / 102
Garden / 104
Intersections / 107
   Masonry bond, Flexible
   intersections, Bond-beam, Brick
Parapet Walls / 111
Retaining / 112

Gravity wall, Cantilever wall,
Counterfort and buttressed
walls
Screen / 120
Serpentine / 121
Veneer / 125
   Steel stud veneer wall, Thin
   veneer

Part Three
Fireplace and Chimney

6 Fireplace / 128

Equipment checklist, Rumford
fireplace, Building code
requirements, Draft, Laying
firebrick, Parts, Planning a
fireplace, Safe operation

7 Chimney / 141

Categories of chimney, Reinforcing
and anchoring, Parts of a
chimney, Laying a chimney,
Troubleshooting

Part Four
The ABCs of Masonry

8 Skills Techniques, and
Procedures / 154

Anchors / 154
   Angle bolts, Expanding

anchors, Adhesive anchors,
Driven anchors, Reanchoring

systems
Apprenticeship Program / 158
Arches / 158
Barbeque and Outdoor / 166
Blueprints, Specifications, and
   Schedules / 167
Bond Break / 175
Bonds, Brick / 175
Bonds, Pattern / 179
   *Square pattern, Herringbone,
   Basketweave*
Caulking / 184
Chases / 186
Cleaning / 187
Coatings and Moisture
   Resistance / 195
   *Preparation, Applying
   cementitious waterproofings,
   Parging, Plaster and stucco,
   Filling a leaking crack, Paint,
   Applying clear coatings, Applying
   membrane waterproofing*
Coloring Mortar / 204
Columns / 205
Concrete Block Construction / 206
Construction Practices / 211
   *Storing material, Laying units,
   Tooling, Cleaning and curing*
Copings / 216
Corners / 217
   *Laying a square lead, Obtuse and
   acute corners, Radial*
Courses, Spacing / 222
Cutting and Sawing / 224
Doors / 228
Flashing / 231
Footings / 233
Glass Block Construction / 236

Grouting / 240
   Preparing, Low lift, High lift
Movement Joint / 244
Jointing / 248
Landscaping / 252
Laying to the Line / 252
Lintels / 253
Modular Planning / 258
Mortar / 258
   *Batching ingredients, Hand
   mixing, Machine mixing, Dry
   batching, Checking the mix,
   Retempering*
Movement, Differential, and
   Cracking / 262
Paving / 262
Piers / 266
Pilasters / 268
Reinforcing / 271
   Joint reinforcement, Placement
Repairs / 272
   *Replacing bricks or blocks,
   Tuckpointing*
Site, Preparation and
   Layout / 274
Steps / 284
   *Terminology, Procedure for
   brick steps*
Stone Construction / 290
   *Laying a veneer wall, Applying
   lightweight artificial stone, Lifting
   stone slabs*
Troubleshooting / 293
Weather / 297
   Cold, Hot
Windows / 304
Vents and Fans / 308

Part Five
Some Fundamentals

9 Estimating Jobs / 309

   *Figuring wall area, Bricks, Mortar,
   Block, Stone, Wall ties and
   reinforcement*

10 Mathematics for Masons / 318

   *Measuring angles, Area, Perimeter
   and volume of shapes, Volume of
   shapes*

11 **Measurements** / 322

United States system, Engineer's measurement, Metric—United States conversions

12 **Safety** / 326

Part Six
Appendices

1 **Building Codes** / 338

2 **Standards for Masonry** / 340

3 **Organizations and Associations** / 341

4 **Glossary** / 344

5 **Index** / 357

INTRODUCTION

Masons are engaged in the oldest building trade. Long before portland cement was invented, masons built the great cities of antiquity. Through the ages, techniques and materials have improved, but the level of skill required has remained high, and masons have always had the satisfaction of knowing that they are building the most attractive and durable structures.

This manual was written to help apprentice and journeymen bricklayers and masons with the many possible applications of the growing number of masonry building products. It will also assist the competent do-it-yourselfer to tackle masonry projects around the home. The manual covers brick, block, and stone work. Footings are described, as are bond beams and other concrete structures that masons must build, but flatwork and poured concrete walls are not.

The alphabetical arrangement and index make it easy to find information on needed topics. Line drawings, charts, and tables describe the important structures and procedures.

Part One describes tools and materials, with the focus on choosing the most appropriate tool or material. Section 1 deals with hand tools, Section 2 with power tools, and Section 3 with scaffolding and ladders. Care and safety for tools and equipment is described. Section 4 details the basic materials—brick, stone, block and mortar. It also describes specialized items—reinforcement, anchors, flashing and moisture-proofing materials. Suggestions for use, common sizes, and specifications are included as appropriate.

Part Two covers the important types of walls: cavity, composite, foundation, garden, etc. Step-by-step procedures and section drawings are included for several walls.

Part Three covers fireplaces and chimneys, describes the many components that can be used in these structures, and suggests how to troubleshoot a fireplace.

Part Four—The ABCs of Masonry—has an alphabetical listing of the most important skills, techniques, and procedures for masonry. The emphasis is on giving step-by-step procedures for everything from using anchors to installing vents. Flexibility is built into the instructions to help meet most of the challenges facing a working mason.

vii

Part Five describes some fundamentals of the building trades, such as estimating jobs, mathematics for masons, measurements, and safety on the job.

Part Six contains information on building codes, standards for masonry, addresses of trade associations and unions, and a glossary of important terms.

I am grateful to Joseph Germ, a veteran mason and instructor of apprentices who contributed greatly to the scope and quality of this book, and to John Thompson, a mason and architect who reviewed the manuscript. I am also indebted to the Brick Institute of America for the use of their graphics.

PART ONE
YOUR TOOLS AND MATERIALS

1

Hand Tools

Masonry has always relied on hand tools, and while some have remained substantially unchanged over the centuries, others are recent inventions to fill needs and boost productivity. Masons must know how to choose the best tool for each job. That objective is the focus of this section.

BRUSHES

Brushes are used for finishing the jointing procedure, for wetting surfaces, cleaning up tools, and applying acid and other cleaning compounds. Do not leave brushes in water while not in use; soaking damages bristles (see Fig. 1.1).

ACID

An acid brush has nylon or fiber bristles. Either short or long handles are available but a long handle is preferable for safety. This kind of brush is used to apply acid for cleaning off mortar, stains, and contaminants.

1

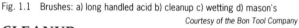

Fig. 1.1   Brushes: a) long handled acid b) cleanup c) wetting d) mason's

Courtesy of the Bon Tool Company

CLEANUP

Cleanup brushes are used to clean tools and equipment after work. They may have short or long handles. When using cleanup brushes, try to distribute wear evenly so the front bristles do not wear out before the rear ones.

MASON'S

With its rather soft bristles, the mason's brush is used to brush mortar burrs from a freshly-laid wall.

WETTING

A wetting brush is used to moisten a surface to assist curing or increase bonding. Keep wetting brushes in good condition, and allow the bristles to dry between uses. Also, hang the brushes from nails to preserve the bristles. Both nylon and natural boar bristles are used. Boar bristles are thought to hold more water and thus have better wetting action.

WIRE

Wire brushes are used to clean hardened mortar from tools and equipment.

CHALK LINE

A chalk line, or chalk box, is a small metal case holding a reel of mason's line dusted with powdered chalk. When the line is removed from the box, the chalk clinging to it is used for marking straight lines during layout. Most chalk lines can also be used as plumb bobs.

CHISEL

Chisels, also called brick sets or bolsters (see Fig. 1.2) are used to score brick and concrete block for cutting. These chisels are sharpened from one side only; the other side remains parallel to the direction of striking. Various widths are available; a 4-inch model is good for general use because it spans the width of most bricks. Strike the set with a mash hammer, not a brick hammer, which is not made to hammer steel. Start with gentle blows and increase their intensity. Make sure the brick or block is uniformly supported

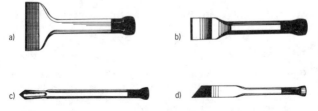

Fig. 1.2   Chisel styles: a) brick set b) mason's chisel c) star chisel d) plugging chisel

Courtesy of the Bon Tool Company

during cutting—otherwise the break may not follow the desired line.

A mason's chisel is used for cutting stone, bricks, and blocks. It may be narrower than a brick chisel.

A star chisel is used to drill holes in masonry material, particularly concrete. Strike it with a mashing hammer and rotate the drill during use.

Tuckpointing, or plugging, chisels have an angled face to aid in removing degraded mortar.

Chisels sharpened from both sides are used for cutting holes, trimming, and other tasks.

Chisels should be sharpened with a grinding wheel. Use water to cool the tip when sharpening. Overheating can ruin the steel's temper, making it brittle. Always wear goggles when using or sharpening chisels.

CUTTER

Wire cutters are used to cut reinforcing wire and mesh. Clean and oil the tool periodically. Heavy-duty cutters are used for cutting re-bar and mesh.

FLOAT

Two types of floats are used in the masonry trade: concrete and rubber.

CONCRETE

Concrete floats are used after striking a slab, to settle the large aggregate and bring up the fine cement–sand paste. The surface of the concrete must be free of standing water before floating is started. Floating is followed by finishing with a trowel. Concrete floats may be made of wood, fiberglass, aluminum, or magnesium. Sizes of hand floats range from 12 inches × 3 inches to 20 inches × 5 inches.

For larger areas, bull floats are used. Bull floats, made of wood, aluminum or magnesium, have extendable handles to allow a worker to float a large surface without walking on it.

RUBBER

Rubber floats are used to smooth parging, to blend repairs to old work, and to grout ceramic and glazed tile. Soft rubber is better for use with soft mortar. Parging must have time to set before floating. Sprinkle some water on the surface and wet the float before beginning. Avoid using too much water, and do not float any section excessively.

HAMMER

Masons use both brick and mash hammers in their work.

BRICK HAMMER

Brick hammers have a square face for striking and a hooked end for starting a cut, trimming burrs, and cleaning mortar from old brick. Several weights are available. A wooden handle deadens vibration better than a steel one but it is not as durable. Use goggles when cutting brick.

MASH HAMMER

A hammer in the 2- to 3-pound range with two round or octagonal faces is called a mash (or mashing) hammer. Do not strike a brick set or other chisel with a brick hammer, which is too hard and usually too light for the purpose. Use a mash hammer instead—it is quicker, safer, and better for the tool.

JOINTER

Several tools are used to compress mortar into the joint and remove excess mortar. Jointers commonly make either a concave or a "vee" joint, although other shapes are available. S-jointers have different shapes or widths on each end for versatility (see Fig. 1.3).

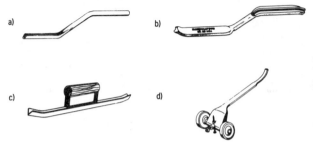

a)   b)   c)   d)

Courtesy of Marshalltown

Fig. 1.3   Jointers: a) s-jointer b) grapevine jointer c) sledrunner d) joint raker

"Sledrunners" are large jointers with wooden handles used to joint large work easily, with minimal damage to the knuckles.

Most jointers are made of steel. Because steel can discolor white and light-colored mortar, plastic, ceramic, stainless steel, or glass jointers are used instead. Good jointers have enough clearance to prevent knuckle injuries while in use.

Joint rakers are wheeled tools used to rake out mortar for installation of movement joints or to create a raked appearance. The depth of raking is adjustable. The wheels must be oiled regularly.

LAYOUT TOOLS

Masons are sometimes required to establish lines of equal height and specified angles for starting excavation, and locating footings and foundations. One of the most common tasks is to "shoot heights" (establish the same height at different points of a site). Three varieties of tripod-mounted tools are used to accurately and efficiently perform these layout tasks: builder's level, transit, and laser transit (see _Site Preparation and Layout_, p. 274).

BUILDER'S LEVEL

A builder's level is used to find relative elevations. The tool consists of a telescope, mounted on a tripod, which can be focused on remote objects. The device is usually carried in a case and affixed to the tripod at the work site. The basic parts of this level are:

- *The telescope, which can range in power from 12 to 32 magnifications, or power. The scope has cross hairs, which are sighted on the distant object.*

- *Spirit levels are used to detect when the tool is perfectly horizontal. This leveling must be done at every location where the level is used.*

- *Leveling screws are used, in conjunction with the spirit levels, to adjust the level to a horizontal position.*

- *The horizontal scale is used to find the angle between the reference stake and the target stake.*

- *A base plate is used on more elaborate levels to simplify the task of leveling the instrument.*

- *The leveling rod is held by an assistant and sighted from the telescope. It is used to determine the difference in height between the telescope and the target stake.*

Builder's levels are precision instruments that must be treated properly and kept in good adjustment. Check the accuracy as often as recommended by the manufacturer. Only experienced technicians should try to adjust a level.

TRANSIT LEVEL

Transit levels use the same principle as builder's levels, but their telescopes also swing in a vertical plane. This gives the tool more versatility, because it can not only shoot heights, but can also be used to establish a row of stakes or to plumb building lines. Some transits use the vernier scale to give extremely accurate angle measurements. Transits incorporate all the features of builder's levels, plus:

- *A compass needle to detect magnetic north.*

- *A compass circle to show the relation of the telescope sightings to magnetic north.*

- *A pivot allowing vertical movement of the telescope.*
- *A vertical scale to show the telescope's vertical angle.*

AUTOMATIC LEVEL

The automatic level is an improvement over the transit and builder's level because it levels itself automatically. After the operator sets it approximately level with manual adjustments, the device levels itself exactly.

LASER TRANSIT

A laser transit emits a beam of coherent light that does not spread appreciably as it travels. The laser is set up on a tripod and leveled. It can be used to measure distance, angles, and heights. A special beam-detecting target is used. A handheld device can read out heights to an accuracy of 0.01 feet.

LASER SAFETY

Lasers can damage unprotected eyes. The following are minimum safety rules; consult OSHA or the manufacturer for more information.

- *Prominent warning signs must be posted in the work area.*
- *Keep unnecessary personnel out of the work area.*
- *The laser should be operated only by qualified personnel. Operators must carry their operator's card at all times.*
- *Manufacturer's instructions must be observed.*
- *Never look at the beam or at any polished surface that might reflect it. Use safety goggles if the beam exceeds 0.005 watts (5 milliwatts).*
- *Never point the beam at anyone. Use a beam height that will put most workers out of range. Place a cap over the beam or shut it off during breaks in work.*
- *Do not allow light intensity exposures above these limits:*

    Direct staring: 1 microwatt per square centimeter.

    Incidental observing: 1 milliwatt per square centimeter.

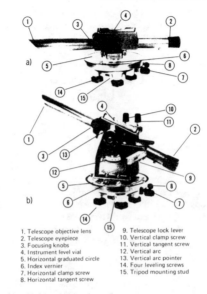

Courtesy of David White Instruments

1. Telescope objective lens
2. Telescope eyepiece
3. Focusing knobs
4. Instrument level vial
5. Horizontal graduated circle
6. Index vernier
7. Horizontal clamp screw
8. Horizontal tangent screw
9. Telescope lock lever
10. Vertical clamp screw
11. Vertical tangent screw
12. Vertical arc
13. Vertical arc pointer
14. Four leveling screws
15. Tripod mounting stud

Fig. 1.4   Parts of level: a) level b) level-transit

LEVELING RODS

A leveling rod is a pole that is usually extendable to 12 feet to 14 feet and graduated in 0.01 feet. Rods are made of wood or fiberglass. A target is affixed to the rod for more accurate work, allowing measurements to thousandths of a foot. Some rods are read by the transit operator, others by the assistant holding the rod.

SELECTING A LEVEL OR TRANSIT

The length of sightings required will determine the power of telescope required. An 18 power scope is used for sightings up to 100 feet; a 26 power scope for sightings up to about 500 feet.

If it is necessary to measure vertical angles, a transit is required. For great accuracy, use an instrument with a vernier scale. A vernier reading to 5 minutes (1/12 degree) will be accurate to within 1/8 inch in 100 feet—and even more accurate at close range. Automatic levels are indicated where the tool will be used frequently, and moved often.

LEVEL

Levels, sometimes called plumb rules, are needed to keep courses horizontal and corners and walls vertical. They are also used for layout (see also *Builder's Level*, p. 7).

LINE

A line level is a small cylinder holding a bubble level. The device is suspended at the midpoint of a mason's line to determine whether the two end points are of equal height. The tool can be used to check leads before laying courses of a wall. The level is only accurate when suspended at the midpoint of the line.

SPIRIT

A spirit level is a rectangular tool with (usually) three pairs of bubble devices to indicate true horizontal and vertical. The spirit level is used to ensure that corners are plumb and level. It is also used as a straight edge when building up corners.

Most masons use a 42-inch or 48-inch spirit level. Shorter models are useful in tight quarters and are easier to handle. Colored liquid in the glasses is easier to read than clear liquid. Be sure to clean splattered mortar from the level and protect the glasses from damage.

Levels are made of aluminum, magnesium, and wood, usually mahogany or a laminate of several woods. Wooden levels should have metal edge protectors, and they need special treatment. Occasionally, treat wooden levels with boiled linseed oil to preserve the wood. Do not get linseed oil on the glass, as this can cause clouding. Hang wooden levels vertically when not in use. Do not hang them in front of a pickup truck's cab window, as the wet wooden levels may warp as they dry.

WATER

Water levels provide an extremely simple, cheap, and accurate way to lay out a site. A rubber hose is fitted with clear glass pipes at each end and filled with water. Because water seeks its own level, the water levels in each pipe are always of equal height. The level is just as accurate as a transit. However, you can have problems because the hose has a tendency to move and there is no fixed point of reference.

LINE AND LINE STRETCHER

A mason's line, generally of braided nylon, is used to keep each course straight and level. The line must be strong enough to stretch tightly between corners and durable enough to take some abuse. The fixtures holding the line must be strong enough to take the strain and not disturb units that are already laid (see Fig. 1.5).

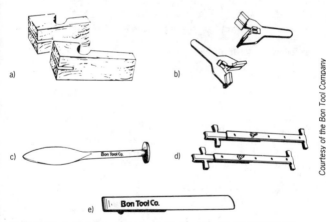

Courtesy of the Bon Tool Company

Fig. 1.5 Lineholders: a) wood b) corner blocks c) line pin d) adjustable line stretcher e) twig

Several means can be used to temporarily attach the line to the corners during construction:

Line stretchers are available in several sizes to fit various sizes of blocks.

Nails or **pegs** are inserted into mortar joints to hold the line. Different styles are available.

Line blocks (or **corner blocks)** hook onto the corner of brick or block.

Trigs (also pronounced **twigs)** are used to prevent the line from sagging or blowing on long courses. The trig can be mounted on a unit that is carefully mortared into the correct position (bonded and plumb) in the course. The line is then slipped inside the clip on the trig.

MISCELLANEOUS TOOLS

BUCKET

Masons use buckets for storing material, measuring ingredients, mixing cleaning solutions, and holding tools. Rubber pails are excellent for their durability and resistance to chemicals.

FILE

Used to dress trowel blades that become dangerously sharp with use.

HOD

A hod is a vee-shaped carrier, closed on one end, used for toting masonry units and mortar on the shoulder. The design makes it easy to unload mortar from the shoulder level. Hods are made of aluminum or plastic.

PLIERS, WRENCHES, AND SCREWDRIVERS

These tools are used for adjusting and repairing equipment and tools. Keeping such tools handy will end time-wasting searches for them.

SHOVELS

Round-bladed shovels are used for measuring sand into the mixer, for cleanup work, and for digging foundations. Flat-bladed shovels are used for loading mortar boards, tempering mortar, and cleanup work.

MASONRY GUIDE (CORNER POLE)

Masonry guides are poles that provide vertical attachments for the mason's line. Most guides are nine feet long so they must be reset for each story. The guide reduces or eliminates the need to plumb each course of the leads. Guides that have standard and modular

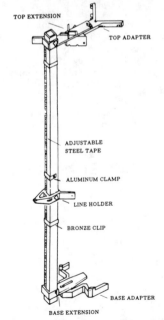

TOP EXTENSION

TOP ADAPTER

ADJUSTABLE
STEEL TAPE

ALUMINUM CLAMP

LINE HOLDER

BRONZE CLIP

BASE ADAPTER

BASE EXTENSION

Fig. 1.6
Masonry guide—
corner pole

Courtesy of the Bon Tool Company

course markings eliminate the need for using brick spacing and modular rules.

The guide is nailed or otherwise attached to the backing wall before brick or block laying begins. Different brackets are used to attach the guide to inside and outside corners, or braces can be used to brace the guide to the ground or floor.

MIXING EQUIPMENT—HAND

Although most masons use power mixers, many carry hand-mixing equipment to prepare small batches of mortar.

MORTAR BOX

A mortar box is a wooden, plastic, or steel box used to mix mortar. The box can have angled sides for easy mixing. It should be built of 2-inch lumber for sturdiness. The box is about 10 inches high; it can range up to 3 feet wide by 7 feet long. A sheet metal lining is useful to prevent wear and water absorption from the mortar. A contractor's wheelbarrow is also useful for mixing.

MORTAR HOE

A mortar hoe is a large, long-handled tool designed to mix mortar quickly. It has two holes in the blade for better mixing action.

MORTAR BOARD OR TUB

A mortar board or tub is used to hold mortar prior to application. Boards are flat and either round or square. A square board is between 24 inches or 30 inches on each side. To lessen absorption of water from the mix, sprinkle wood mortar boards with water before loading it with mortar. Wood boards can be sealed to prevent absorption and simplify cleaning.

Mortar tubs are tapered and about 15 inches square on the bottom and 30 inches square on the top. They are about 6 or 7 inches deep and designed for easy access to the mortar. Both steel and plastic tubs are available.

PLUMB BOB

A plumb bob is a cylindrical, cone-shaped metal tool with a point on the lower end and a hole at the top center. The tool is used to find exact vertical alignment during layout, such as locating a transit or builder's level directly above a reference stake.

RULERS

Masons use a variety of rules to lay out jobs and control the spacing of courses.

BRICK SPACING AND MODULAR

Brick spacing and modular rules are marked with inches on one side and a course pattern on the other. They are both used to achieve the desired spacing of brick and block courses. Both folding and retractable rules are made with this marking. Folding rules have the advantage of staying straight and true when in use, but they are more fragile than retractable rules. Clean and oil joints on all folding rules regularly. Keep them folded for protection.

A modular rule is used for laying modular units, in which a certain number of courses equals 16 inches in height.

Courses per 16" height	Brick or types
2	8" Concrete block
3	Jumbo utility triple
4	Jumbo utility
5	Engineer
6	Standard
8	Roman

Brick spacing rules are used for spacing course heights with standard brick. Each spacing equals the height of one brick plus one bed joint. Ten spacing ranges are given, ranging from 1 (2-3/8 inches high) through 0 (3 inches high). The spacing numbers are marked in red. The number of courses at that height is marked in black alongside the spacing number.

To find the height of 16 courses of #4 spacing, simply search the rule for the one place where a red 4 is alongside a black 16. Then turn the rule over and read the height in feet and inches. To find how many courses of a certain spacing is closest to a desired height, read that height on the rule, then flip the rule over and look for the closest spacing that will work with your units. Adjust the spacing slightly if needed to reach the proper height (see *Courses, spacing*, p. 222).

STEEL TAPE

Two types of steel tapes are used by masons. A 12-foot to 25-foot automatic retracting tape is used for general measuring purposes. A longer, windup tape is used for laying out building sites. Wipe the tape occasionally to clean it. Light oil can be used to aid blade retraction.

SAFETY EQUIPMENT

With increasing attention to safety, masons are using a variety of safety equipment (see also *Safety* in Part Five, p. 326).

GLOVES

Rubber gloves are used to protect the hands during wall-washing procedures. Long sleeves are desirable. Choose a high-quality glove to resist the abrasion of masonry materials.

Protective gloves can be used when handling rough masonry units and mortar. Some masons prefer wrapping their fingers with tape to prevent abrasion.

GOGGLES

Approved eye protection is essential when performing washing operations and when hand-cutting, sawing, or grinding. Wear eye protection when striking steel upon steel, cleaning out cracks, or in any situation likely to create dust, flying chips, or hazardous chemical spray.

HARD HAT

A steel or plastic hard hat is required whenever you are working under other masons or near dangerous building operations.

KNEE PADS

Rubber or leather kneepads are vital to protect knees while kneeling. These pads are especially useful for working on hearths and paving.

STEEL-TOED SHOES

Steel toes are essential to protect toes from falling tools and material. Since cement dries out leather, keep leather oiled and flexible for greater foot comfort and longer shoe life.

SQUARE

A steel carpenter's square is handy for laying up square corners. A square is also used during dry layout of leads and for squaring chimneys.

STONE TOOLS

Special tools are used for working stone. Steel wedges are used to split stones and wooden, lead, or plastic wedges are used as shims during laying. Stone mason's hammers are made in weights from 3 to 16 pounds for different applications. Chipping chisels with blades ranging from 1/4 inch to 2 inches wide are used to shape stone. Toothed and scoring chisels are used for scoring soft stone before cutting (do not use on hard stone or cement). Stone chisels are available with tungsten carbide blades for long life (see Fig. 1.7).

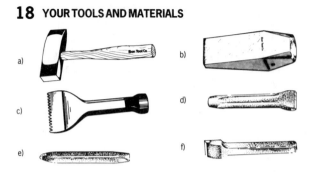

Fig. 1.7 Stone tools: a) stone mason's hammer b) stone wedge c) toothed chisel d) tracing tool e) stone point f) hand chipper

STORY POLE

Story poles are used to check the height of courses during layup. They can be made from 1 inch × 2 inch lumber. It is convenient to mark the height of special courses, such as sills and lintels, to prevent errors. Because masonry guides perform all the functions of the story pole and also align the corners, they may be used instead.

TONGS, BRICK AND BLOCK

Brick tongs are used to carry up to 10 bricks in one hand, saving injury and leaving the other hand free. Most tongs are adjustable to different size loads. The weight of the bricks creates the clamping action and ensures a secure grip.

Block tongs carry a single block in one hand. The force of gravity creates the clamping action.

TROWEL

A variety of trowels are used in masonry. Quality is vital, as some trowels are used all day long. A good trowel is forged from a single piece of tool or spring steel and solidly bonded to a wood or plastic handle. The trowel must have a bright finish to prevent mortar from sticking to it.

Trowels must remain bright and smooth. Scour off dry mortar with sand, a piece of concrete block, another trowel, or a wire brush. Wash and scrub trowels after work. To prevent rust, which will interfere with a smooth mortar delivery, allow the trowel to dry before storing it. Oil trowels if they will be out of service for more than a week (see Fig. 1.8).

TROWEL USE

- *Hold the trowel in a firm but relaxed grip.*
- *Work with both hands so you need not put the trowel down very often.*
- *Keep the thumb and fingers back from the blade so they stay out of the mortar.*
- *Look for ways to speed up your work without sacrificing quality.*
- *Remove dried mortar frequently during the day.*
- *Use care if the trowel has sharp edges.*
- *Use the trowel to line up edges and faces of brick while laying.*
- *When buttering, "settle" the mortar on the trowel after picking it up. Do this by shaking the trowel gently up and down to make mortar adhere to the steel. Do not settle the mortar when spreading bed joints.*

BRICK

The brick trowel is the primary tool for most masons, so it must be chosen carefully. Several patterns are available. Blades range from 10 inches to 12 inches long by 4-3/4 inches to 5-7/8 inches wide.

A Philadelphia pattern brick trowel has a broad heel, while a London pattern is more diamond-shaped. A narrow brick trowel is handy when laying veneer walls, because it is less likely to interfere with the backing and line.

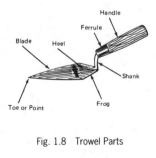

Fig. 1.8   Trowel Parts

BUTTERING

A buttering trowel is a stubby tool sometimes used to butter ends of bricks, particularly when two masons are working in tandem. It is also used wherever a smaller trowel is needed, as in laying firebrick.

CAULKING

Also called a tuckpointing trowel or filler, a caulking trowel is long and narrow. The tool is designed to force mortar into joints during tuckpointing. The blade is about 6 inches long; the width ranges from 1/4 inch to 1 inch. A caulking trowel is also used for striking when a flat joint is desired.

CEMENT FINISHING

A cement finishing trowel is used to create the final surface on concrete. Although most cement finishing is done by cement finishers, masons occasionally prepare small areas of concrete for inner hearths and chimney caps. A finishing trowel may also used for parging. The blade is made of spring steel and can get quite sharp with use. File the edges occasionally to dress them. Sizes range from 12 inches × 3 inches to 20 inches × 5 inches.

MARGIN

A margin trowel has a rectangular blade roughly 5 inches long by 2 inches wide. Margin trowels are useful for mixing small batches of material, cleaning pails and pointing stonework.

PLASTERING

A plastering trowel has a rectangular blade approximately 4 inches by 12 inches and is used for parging.

POINTING

A pointing trowel is a small version of the brick trowel used to point stone and brick.

SWIMMING POOL

A swimming pool trowel is rectangular but has rounded corners. The tool is used to plaster irregular surfaces, such as pools and tanks (see Fig. 1.9).

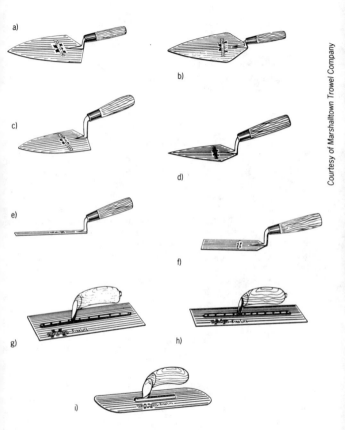

Courtesy of *Marshalltown Trowel Company*

Fig. 1.9   Trowel styles: a) Philadelphia pattern b) London pattern c) tile setter's d) pointing e) tuckpointer f) margin trowel g) plastering h) cement finishing i) swimming pool

2
Power Tools

Modern masons depend on an assortment of power tools to multiply their force, which is essential for dealing with heavy, hard materials. Knowing which power tools to buy, and how to use them safely and effectively, is a trademark of the professional mason.

ELECTRIC CORD

A 50-foot or 100-foot cord is used to run electrically powered equipment in the field. Power draw is measured in amps, and you should make sure the cord has enough capacity for the tool. The longer the cord, the larger it must be to supply proper amperage. Because higher-voltage current allows more power to flow through a wire, some electric tools use 220 volts instead of the standard 110 volts. Make sure to supply tools with the correct voltage.

Safety: Use a cord large enough for the job. Use only grounded cords. Be extremely careful when working in damp areas. Inspect the cord every day before use. Flat cords are not OSHA approved. Cords should have a ground fault interrupter for safety (see *Safety, Electricity*, p. 329).

DRILL

Masons must drill through concrete and masonry material to install anchors and dowels. Due to the hardness of masonry, hammer drills, which create a rapid series of impacts during the drilling, are preferable. The hammering action can usually be turned off if a conventional drill is desired. Hammer drills require special bits; do not attempt to use regular masonry bits (except in the smaller sizes).

A "dead-man" switch is required on drills used at construction sites. These switches have no locking device, so finger pressure is needed while the drill operates; the tool turns off automatically when pressure is released.

Drilling is most productive with sharp bits. Replace dull bits. Most masons use bits with tungsten carbide cutting tips, which remain sharper longer.

Safety: Check tool condition before use. The cord, bit and handles should be in good condition. Use safety glasses and a dust mask or respirator when drilling. Make sure the extension cord is in good shape.

FORKLIFT

Both conventional and extended forklifts are used to hoist mortar and masonry units to masons on scaffolds. Extended lift machines can hoist materials directly to a scaffold even if the machine cannot drive to the scaffold base.

Safety: Forklifts can be dangerous, and should be operated only by a trained operator. Inspect brakes, overhead protection, and all other items specified by the manufacturer each day before using. Service the oil, water, and hydraulic system as required by the manufacturer. Wear the safety belt. Do not overload the machine. Do not make sudden turns. To prevent tipovers, back down a ramp. Go slowly across rough terrain. Solid tire trucks are to be used only on pavement, never in the field. When operating fuel-powered lifts in an enclosed area, provide adequate ventilation and beware of carbon monoxide poisoning.

GRINDER

Grinders are used both to remove mortar before tuckpointing and as cutoff tools. Make sure the wheel is appropriate for the material to be worked; check with the manufacturer if necessary. Never use a damaged grinding wheel. Do not remove shields. Eye and ear protection is required; a respirator is suggested, especially in enclosed areas.

HOIST

Hoists are used for lifting materials to the scaffold. They should not be used for personnel, and should be operated by a qualified operator.

Safety: Inspect daily prior to use. Do not overload. Maintain according to manufacturer's specifications.

MORTAR MIXER

Gasoline and electric mixers are used for all but the smallest masonry operations. Capacities range from 3 cubic feet to 9 cubic feet. Three cubic feet of mortar can be made from one bag of masonry cement and the appropriate amount of sand and water.

For safety, most mixers can be charged and discharged without removing the mixing drum guard. Do not remove any guards while the motor is running. The mixer should be cleaned after use, with the motor off and the gears disengaged.

Handle gasoline carefully and store it only in an approved container (see *Safety, Flammable liquids*, p. 331). Use a respirator when handling cement. Ground electric mixers to prevent electrical shock.

POWER WASHER

Power washers are used to wash walls after construction and old masonry structures during renovation. Observe electrical safety precautions and any cautions applicable to the cleaning compound being used. Use proper gloves, masks, and other protective clothing as required.

SANDBLASTER

Sandblasting is a very effective means of cleaning almost every sort of stain and dirt from the surface of masonry, but it damages many surfaces. Several types of treatment are possible. Rounded, urn sand can be used to reduce surface damage. Wet sand is used to remove surface coatings such as paint. Silica-free aggregate and low-pressure water are appropriate for soft masonry and ornamental features. Test sandblasting on an inconspicuous area to find the optimum angle and distance from the wall. Inspect for damage before proceeding.

SAW

Power saws are used to cut units either on the ground or scaffold. Units are placed on a carriage and run through the blade, which spins on a horizontal axis and swings down into the cut. The carriage may have a miter gauge for easy cutting of mitered units.

Stationary saws sit on the ground; portable saws can be used on the scaffold or indoors. A third variety is the hand-held saw that can be moved to wherever a cut must be made. Some saws are gasoline powered, but most are electric, usually running on 220 volts. Some saws will operate on either 110 or 220 volts.

Blades can be carborundum or diamond. Many saws have a cooling apparatus that pumps water to the cut to cool the blade and reduce dust. Refer to manufacturer, dealer, or a company representative for advice about the best blade for the material you plan to cut.

Safety: Only a trained operator should attempt to use a masonry saw. Check the equipment before using. Check the voltage before connecting the saw to a power source. Make sure all guards are in place and that the saw is stable. Observe gasoline safety precautions with gasoline-powered saws. Use the correct blade and rate of feed. Wear goggles, respirator, and ear protection. Never allow a wet saw to run dry (see *Safety, Electricity,* p. 329; and *Safety, Flammable liquids,* p. 331).

STUD TOOL

A stud tool uses powder charges to insert hardened nails or studs into concrete, masonry, and steel. The tool can be used to nail a furring strip to a masonry wall or to drive threaded studs into masonry for attaching various building elements. Stud tools are probably the fastest means of fastening to masonry, although the charges and fasteners make them more expensive than many other systems. The tools are dangerous and must be used only by a trained operator.

The charge is inserted into one end of a piston, and the nail or stud set into the other end. When struck with a hammer weighing about 3 pounds, or when the trigger is pulled, the charge explodes, driving the nail or stud into the base material.

Some tools have clips to feed charges quickly. Different strength charges are available for fastening to different materials, and to various depths.

Safety: Low-velocity charges are preferable because they limit the flight of the stud or nail in case of accident. The operator should be licensed by the distributor and thoroughly familiar with tool safety, maintenance, and operation. Safety glasses, face shield, and ear protection should be worn. Observe other procedures during use. Inspect and test the tool each day and correct any defects before use. Do not load the tool until immediately before use.

WHEELBARROW AND CARTS

A wheelbarrow is used for mixing mortar and transporting mortar and other materials across rough ground. Most wheelbarrows have a pneumatic tire for easy wheeling. Capacity ranges from 4-1/2 to 6 cubic feet. The wheelbarrow should be cleaned after holding mortar. Check tire pressure occasionally and grease bearings if needed.

Four-wheeled buggies are used to transport mortar and concrete across uneven ground. Both pneumatic and solid tires are available.

Brick and block carts are used to transport units around the site. They may be used on scaffolds as well as the ground.

3

Scaffolding and Ladders

As soon as a masonry job reaches chest height, some form of scaffolding is needed, as it is impractical to spread mortar and lay units above this height. Horses and planks are acceptable for low scaffolds, but any substantial job requires steel or aluminum scaffolding.

A scaffold must be strong enough to hold the workers and masonry material; it must be safe enough to prevent workers from stumbling and falling, and it must be convenient to erect and dismantle. It should allow free movement by the masons and their helpers. Two levels of scaffold are required. Masons walk on the lower level, which is nearer the wall than the upper level. The upper level holds mortar and masonry units. Guards must be installed as required by safety regulations.

Scaffolding must have provision for hoists or access by forklifts. Planks on some scaffolds must be raised periodically. On other scaffolds, the entire working area can be lifted gradually with no need to stop work for readjustment.

A guardrail is required on all scaffolds above 4 feet high, unless the platform is more than 45 inches wide. Guardrails and toe boards are required for any platform above 11 feet high.

Safety: Scaffold safety is a complicated topic. Some basic safety information is given for most types of scaffold described below. For detailed safety information, consult the manufacturer, the Scaffolding, Shoring and Forming Institute, or OSHA (see *Safety, Walking and Working on Surfaces*, p. 335).

GENERAL SCAFFOLD SAFETY

• *Inspect all equipment each day before use. Do not use any defective components.*

- *Assemble the scaffold properly, following manufacturer's directions. Take no shortcuts.*

- *All footings and attachments must be sound and secure.*

- *Do not overload or use a scaffold for purposes for which it was not designed.*

- *Avoid electric power lines.*

- *Observe safe work practices on and beneath the scaffold.*

BOATSWAIN'S CHAIR

A boatswain's chair is a sling designed to suspend a worker from above. It is handy for small repair jobs or inspecting buildings, but not for extensive jobs.

Safety: Inspect the equipment before using. Use a safety line with a separate attachment to the structure. Do not overreach or overload. Avoid electric wires. Do not use in high wind.

CHIMNEY SCAFFOLD

A chimney scaffold can be built on the job, but ready-made units are also available. The scaffold should rest on the ridge top if possible. If the chimney is not near a ridge, nail the scaffold if the roof has not been laid. Otherwise, fasten the scaffold to planks that extend beyond the ridge and are connected to planks sloping down the opposite side.

Safety: Use a safety line on roofs. Do not overload the scaffold. Inspect scaffold before use. Do not alter the position of the scaffold until all tools and material are removed. Use a roof ladder on steep roofs.

HORSES AND BUCKS

Horses and sawbucks are still used for low scaffolding for jobs, such as veneering single-story houses.

Safety: Make sure all bucks are solidly based on level ground. Inspect all equipment before use. Avoid overloading and overreaching.

LADDER

Ladders are used on job sites to provide access to scaffolds and other high places. Wood, metal, and fiberglass ladders are all acceptable if they conform to OSHA standards and are inspected before and during use. Wood or fiberglass ladders must be used if electric wires are nearby.

A roof ladder is useful for ascending steep roofs for working on chimneys that are far from the roof edge. The ladder must be securely built and have a good grip on the ridge of the roof.

Safety: Do not use a ladder as a working platform. Careful setup and use are vital to safety. Discard or repair ladders showing any signs of wear, especially those with broken or split rungs or side rails. Portable ladders must have clear access at bottom and top. Job-made ladders must have cleats inset 1/2 inch into the side rails or held in place with filler blocks. Cleats must be uniformly spaced 12 inches top to top.

Setup: Ladder feet should be set about one-quarter of the working height away from the wall. The feet must be solidly footed. On smooth surfaces, prevent slippage by using treaded feet or rest the feet against a cleat nailed to the floor. Tie or otherwise secure the ladder top to prevent movement. If possible, remove the rungs above the landing to simplify stepping on and off the landing. Allow the ladder to extend about 3 feet above the landing so workers have a hand grip while stepping on and off the ladder.

Use: Inspect the ladder before use and frequently during use. Clean shoes before climbing. Clean and dry wet or slippery rungs. Face the ladder and climb with both hands. Do not carry any tools or equipment up the ladder if a hoist is available. No horseplay on the ladder; allow only one person at a time on the ladder.

ROLLING TOWER

Mobile towers can be rolled along a smooth, flat surface without the need for disassembly and reassembly, but they should only be used for jobs near good paving (see Fig. 3.1).

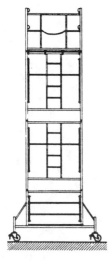

Fig. 3.1
Rolling tower

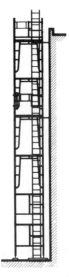

Courtesy of Safway Steel Products

Fig. 3.2
Scaffold side view

MORGAN TOWER

A Morgan tower is an adjustable platform that is raised by winches as the masons proceed. The device increases efficiency because there is no need to work at uncomfortable heights or to stop to raise scaffolding.

SWING STAGE

Swing stages are suspended from parapet walls to support workers. They are commonly used to repair existing buildings and wash windows.

Safety: Swing stages can be dangerous. The following are just a few guidelines for safety. Consult the manufacturer or OSHA for detailed information:

- *Always wear a safety harness that is independently secured to the building.*
- *Inspect daily all equipment, including wire ropes, clamps, or other attachments at the roof.*
- *Erect the scaffold carefully in accordance with manufacturer's instructions.*
- *Secure tiebacks of adequate strength to structurally sound building components.*
- *Do not overload.*
- *Avoid power lines or have the utility turn the power off.*
- *Allow the bottom end of wire ropes to hang freely.*
- *Ensure that all guardrails, platforms, and toe boards are properly installed.*
- *Do not use a swing stage unless you have personally checked the roof support system and all other safety precautions have been checked.*

TUBULAR

Tubular scaffold, also called sectional scaffolding, is a versatile system that uses sections of steel stacked on top of one another and braced with diagonal braces. Planks are placed on the front crosspieces for workers and on the rear crosspieces for materials (see Fig. 3.2).

Many options are available:

- *Brackets allow masons to work closer to the wall.*
- *Different widths and heights of standards.*
- *Some sections permit pedestrians to walk underneath the scaffold and permit laborers to easily walk around the scaffold.*
- *Screwjacks provide sure, safe adjustments for uneven footing surfaces.*
- *Internal stairways eliminate the need for ladders.*
- *The scaffold can be converted into a rolling tower.*

4
Materials

Masons deal with two basic classes of building materials: fired clay products and portland cement products. Fired clay includes bricks, some paving units, and many forms of tile, including flue, structural, and facing tile. Portland cement forms the bonding agent in concrete, concrete block, and is combined with lime and sand to form masonry cement.

A wide variety of other materials is used to increase the strength, water resistance, appearance, and flexibility of masonry structures. These include anchors, reinforcement, flashing, coatings, and impregnating compounds. Masons must know how to select these materials in the most economical and effective manner.

ANCHORS

Anchors are fasteners that are installed either during or after construction. Angle bolts are bonded into masonry or concrete during construction. Post-construction fasteners are usually inserted into holes drilled in masonry, where they expand to create the fastening action. Some anchors bond to the masonry with adhesives; others make their own hole and bond by friction.

Applications for anchors include fastening material to existing structures and repairing cavity walls that have separated. Each manufacturer has a different array of anchors; consult their literature for details on products.

Other steel products used in the trade are sometimes called anchors. See *Reinforcing and Ties*, p. 81 for information on joint reinforcement, wall ties, and other steel products used in masonry construction.

Anchors are subjected to two types of force: tension and shear. Tension pulls the anchor in line with its axis, perpendicular to the surface of the base material. Shear tends to snap off the bolt head parallel to the surface of the base material (see Fig. 4.1).

Anchor strength is determined by anchor style, diameter and length, penetration, condition, and type of base material, and method of engagement.

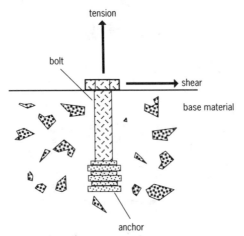

Fig. 4.1 Anchor stresses

SPECIFYING ANCHORS

Consider five factors when choosing anchors:

Load requirement. The total weight of the object to be fastened divided by the number of anchors to be used equals the load requirement per fastener. For safety, use an anchor rated at four times that amount.

Load condition. Four types of load conditions affect the choice of fastener:

1. Dead loads are subject to no outside force.

2. Variable loads change gradually over time.

3. Vibratory loads are subject to high-frequency impact, such as from a motor.

4. Shock loads are subject to shock at uncertain intervals, such as guardrails.

Base material. Fasteners must be suitable to the base material.

- *Solid concrete presents few problems, and the anchor's holding power depends directly on the compressive strength of the concrete.*

- *Aged concrete may chip. Use an anchor that will adapt to an uneven hole.*

- *Hollow concrete (such as precast hollow-core slabs) must be treated carefully per manufacturer's instructions.*

- *Blocks usually have low compressive strength and require an anchor with high expansion characteristics.*

- *Hollow walls may be fastened with wing or toggle bolts that expand after being inserted into a hole.*

- *Stone is problematic because it has variable hardness and tends to crack.*

- *Brick and mortar have varying degrees of hardness. In hard brick, low-torque and/or low impact anchors are desired. For soft brick, use a draw-up anchor or one that is caulked in place. For mortar joints, use a device that fits into the joint and engages both adjacent bricks.*

Atmospheric condition. Plated steel resists rust but the plating may be rubbed off during installation. Stainless steel, lead, and zinc anchors are available for various situations and chemical environments.

Cost. Cost is determined by fastener and labor cost, loss rate, and equipment required. Commonly, more expensive anchors are easier and faster to set than less expensive ones.

ANCHOR TYPES

Anchors are made in great number of styles, sizes, head types, lengths, diameters, and finishes. Anchors fall into five general types:

1. Angle bolts have a bend at one end and are used to tie sills and plates to masonry walls and to suspend joists. The bent end can have a 90° or greater bend, or it can have a hook shape. Angle bolts are usually 3/8 inch or 1/2 inch in diameter and up to 15 inches long. They are grouted into block cavities, mortared into composite walls, or cast into concrete, at the required location.

2. Anchors that expand into the hole. The three basic types are:

a. Toggle bolts employ a wing-like mechanism that retracts to fit through a hole and expands once through the hole. Toggle bolts are good for hollow units, but not for solid masonry. They may be removed but not reused because the toggle usually falls off during removal.

TABLE 4.1

Fastener Selection Chart

Fastener	Brick Type			Installation Location			Fixture Weight		
	Solid Brick (Cored)	Solid Brick (Uncored)	Hollow Brick	Head Joint	Bed Joint	Unit Face	Light	Medium	Heavy
Wooden Blocks	✓	✓	✓	✓			✓	✓	
Metal Wall Plugs	✓	✓	✓	✓	✓		✓	✓	
Screw Shields and Plugs	✓	✓	✓	✓	✓		✓		
Toggle Bolts			✓		✓	✓	✓	✓	✓
Hollow Wall Screws			✓		✓	✓	✓	✓	
Screws	✓	✓	✓	✓	✓	✓	✓	✓	
Sleeve Anchors	✓	✓	✓	✓	✓	✓	✓	✓	✓
Wedge Anchors	✓	✓		✓	✓		✓	✓	✓
Lag Shields	✓	✓		✓	✓		✓	✓	✓
Masonry Nails	✓	✓	✓	✓	✓		✓	✓	
Powder-Driven Fasteners	✓	✓		✓	✓		✓	✓	
Adhesives	✓	✓	✓	Surface Applied			✓		

b. Wedge anchors fit into a hole and expand upon tightening. Most work by drawing a cone-shaped lug toward the bolt head. This lug expands flaps which are forced to the sides of the hole, creating the fastening action and a great deal of stress. Wedge anchors are better suited to solid concrete than unit masonry. A torque wrench may be used to properly tighten the anchor without damaging the base material.

c. Bolts and shields (also called sleeves) also expand into a hole. Shields are made of rubber, nylon, fiber, lead, or plastic. When the bolt is screwed into the shield, the shield expands to grip the sides of the hole. Some two-piece anchors with machine threads are removable and replaceable. During installation, check that the expansion device or shield is firmly seated in solid masonry material. Shields and cinch bolts must not turn in the hole or they will not tighten.

3. Adhesive anchors are cemented or epoxied into a drill hole to create a chemical and physical bond to the base material. When used in mortar, sound concrete or masonry, these anchors are suited to heavy-duty and impact applications, such as anchoring machinery, drive equipment, parking meters, fences, railings, motors, pulleys, and bolts.

Adhesives may be used for fastening electrical boxes, furring, and paneling. The material must be chosen according to material in the wall and attachment, and chemical, weather, and temperature conditions at the location. Follow the manufacturer's directions when using any adhesive.

4. Anchors that are driven into a surface achieve their bond by friction with the sides of the hole. The two types are:

a. Masonry nails are case-hardened, with either a spiral or a cut shape. Nails should be hammered into mortar, not into units. In single-wythe, exterior walls, nails can create cracks through which water can enter.

b. Powder-actuated studs and nails. Both nails and threaded studs are available for these tools. Charges range from .22 to .38 caliber. Charges can be chosen to match the fastener size, and depth and density of material. Powder-actuated anchors are often the quickest means of anchoring, though they require special tools and training. At first, the failure rate may be high as the operator adjusts to the base material and nail size. Tools should only be operated by trained personnel wearing appropriate safety equipment (see Figs. 4.2 and 4.3).

5. Reanchoring systems are used to strengthen older two-wythe walls that suffer from any of these problems: a lack of masonry bonders or wall ties; existing ties or bonders that have failed or don't meet current codes; or a new veneer is needed over an existing structure.

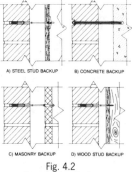

A) STEEL STUD BACKUP B) CONCRETE BACKUP

C) MASONRY BACKUP D) WOOD STUD BACKUP

Fig. 4.2
Reanchoring Systems

Courtesy of Brick Institute of America

The three general types of reanchoring systems are: mechanical expansion (either screw and shield or toggle type); adhesive; and screw. Mechanical expansion and adhesive types are suited to two-wythe walls. Screw types are suited for veneers backed up with wood or steel studs. Adhesive anchors for cavity walls incorporate a wire mesh shield which prevents the epoxy from escaping the drill hole during installation.

BLOCK, CONCRETE

Concrete blocks are a sturdy, economical means of constructing walls. They are used in many applications requiring fire resistance and compressive strength. Properly constructed, block has great weather resistance and extreme durability.

First cast in 1882, concrete blocks have become a common and useful building material. Between the United States and Canada,

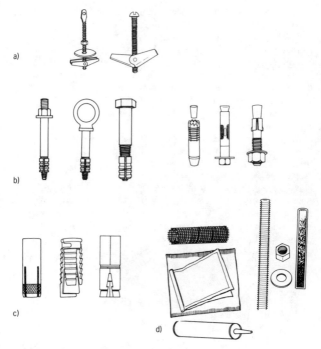

Courtesy of U.S.E. Diamond, Inc.

Fig. 4.3 Expanding anchor styles a) toggle bolt b) cinch bolt c) lag shields
d) adhesive anchor

the equivalent of about 4 billion $8 \times 8 \times 16$ units are produced annually. Concrete blocks are formed and cured under controlled conditions. A concrete of portland cement, graded aggregate, and a minimal amount of water is molded into the desired shape. Curing is begun in a steam-heated oven, which reduces shrinkage drying by about one-third. After heating, the units are cured and dried for the required period before shipment. The mix and processing techniques can be varied to meet product requirements. Proper mix and processing are necessary for strength, color, sizing, and texture.

New concrete products shrink as they dry and then expand if they gain moisture, but a wall will never be larger than the day it was laid. The contraction is more likely to cause problems in a building, but both expansion and contraction must be taken into account when designing a concrete block structure. This is generally done by providing control joints in areas likely to crack.

A wide variety of shapes and sizes are available; when laid with a 3/8-inch joint, the nominal sizes fit the 8-inch module. Half-height blocks are available in many styles, and concrete bricks and other solid units are also made (see Fig. 4.4). Block manufacturers will produce a run of a custom shape for a mold fee and a minimum order (see Fig. 4.4).

Blocks are available in heavy and light weights. The density of heavy concrete block ranges from 130 to 145 pounds per cubic foot of material. (Because most blocks are hollow, actual unit weight is less.) Lightweight blocks are made of materials such as cinders, pumice, and expanded shale.

Compared to heavyweight blocks, lightweight blocks:

- *Are 20 to 40 percent less dense.*
- *Insulate better.*
- *Have similar compressive strength (in most cases).*
- *Are more expensive.*
- *Tend to shrink more*

Colored blocks are a welcome change from portland cement grey. Tan, buff, red, brown, pink, black, and yellow are commonly sold. Blue, if available, is likely to fade. Dark green is expensive. Colored blocks, like all others, should be covered on the job site. If blocks are not uniform in color, distribute them to various locations rather than placing all blocks of a particular hue in one part of the wall.

Block textures vary from smooth to rough. Two types approximate the appearance of natural stone: slump block and split block. Ribbed block, formed by splitting a double unit when partly cured, is becoming more popular. The parts of blocks have their own nomenclature: the outer sides are shells, the members joining the shells are webs, and the hollow spaces are cells, or cores.

Regular stretcher

One plain end (single corner)

Both ends plain (double corner or pier)

Slot for breaking

(a) Two-core 8 x 8 x 16-in. units

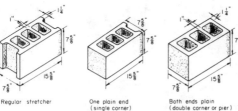

Regular stretcher

One plain end (single corner)

Both ends plain (double corner or pier)

(b) Three-core 8 x 8 x 16-in. units

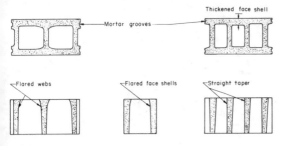

Mortar grooves

Thickened face shell

Flared webs

Flared face shells

Straight taper

(c) Cross sections

Courtesy of Portland Cement Association

Fig. 4.4   Concrete block styles

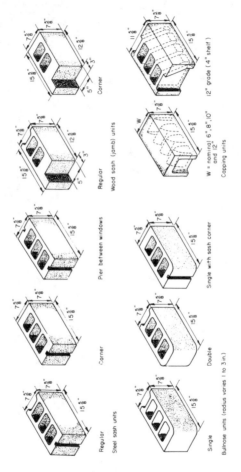

Fig. 4.4 Concrete block styles (cont.)

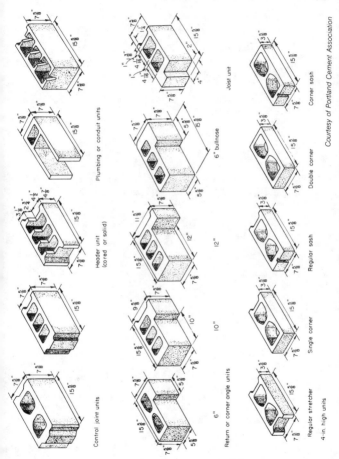

Courtesy of Portland Cement Association

Plumbing or conduit units

Header unit (cored or solid)

Control joint units

Joist unit

6" bullnose

12"

10"

6"

Return or corner angle units

Corner sash

Double corner

Regular sash

Single corner

Regular stretcher

4-in high units

Fig. 4.4   Concrete block styles (cont.)

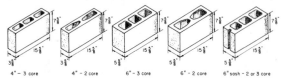

4" – 3 core   4" – 2 core   6" – 3 core   6" – 2 core   6" sash – 2 or 3 core

4 and 6-in. partition and backup units

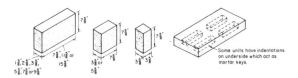

(b) Solid units

(c) Cap or paving unit

Some units have indentations on underside which act as mortar keys.

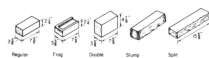

Regular   Frog   Double   Slump   Split   Hollow-perforated

(d) Concrete brick

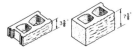

(e) Slump block

(f) Split-face units

(g) Split block yeilding two units

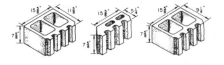

(h) Ribbed split-face units

Courtesy of Portland Cement Association

Fig. 4.4   Concrete block styles (cont.)

Blocks are seldom shipped very far, and the availability of lightweight block, special shapes, and color, varies by location.

Blocks have varying degrees of absorbency. ASTM standards require that blocks absorb no more than 15 pounds of water per cubic foot of concrete. When laid, units must contain no more than 40 percent of this maximum, because blocks that are laid too wet will not bond properly and will contract and possibly crack the wall.

Liquid polymeric admixes have been developed to render block virtually waterproof. The admixes have helped industry to produce slump and split blocks, which would otherwise admit too much water. The admixes also help the blocks retain their color. The manufacturer suggests that a companion admix be added to mortar during mixing (see *Mortar, Ingredients, Admixtures*, p. 71).

BLOCK, CHIMNEY

Chimney blocks are an economical and rapid means of building a chimney, although they are not as attractive as a brick chimney. Blocks are available to accommodate 8 × 8 and 8 × 12 flues. Chimney blocks are not recommended for exterior use in areas subject to freezing temperatures.

Some suppliers sell special units for clean-outs and thimbles. Otherwise, these details are made by sawing chimney blocks or building the requisite courses with other concrete masonry units.

BLOCK, GLASS

Glass blocks are non-loadbearing masonry units made by molding two sections of glass and bonding them together. The interior is hermetically sealed to prevent condensation. Most glass blocks allow light to pass through but distort the view to ensure privacy. Relatively transparent blocks are also available. Special blocks are designed to direct light upward or downward for certain building situations and bulletproof blocks are available.

With the recommended 1/4-inch mortar joint, square glass blocks have nominal sizes of 6 inches, 8 inches, and 12 inches. New, rounded-end blocks are available to cap the sides or tops of block walls. The most common size of glass block is 7-3/4 inches square by 3-7/8 inches thick, to fit the 8-inch modular system. Other sizes and shapes are available.

The mortar-bearing surfaces of glass blocks are roughened to accept mortar. An expansion strip is required between glass block and adjacent material. Liberal use of joint reinforcement is suggested for glass block walls. Partition walls using rounded-end blocks must have support on three sides of the panel.

Metal or plastic spacers shaped like a plus sign can be used to ensure accurate spacing between glass blocks. Glass blocks can be used to create curving walls, with the minimum radius dependent on the size of unit being used.

TABLE 4.2
Curved panel radius for glass block

Inside joint = $\frac{1}{8}$ inch		Outside joint = $\frac{5}{8}$ inch
Block size (inches)	**Blocks per 90° arc**	**Outside radius (inches)**
4×8	13	35.8
6×6	13	52.4
8×8	13	69.0
12×12	13	102.1

FURNISHED BY GLAUSHAUS, INC., DISTRIBUTORS WECK GLASS BLOCK.

BRICK

Bricks are the oldest masonry building units. Early bricks were made by many cultures of the Middle East. At first, dried clay, similar to today's adobe, was used. As builders realized that fired clay was stronger and more durable, bricks as we know them came into being.

Modern bricks are made of clay or shale that has been fired to about 1950°F (1065°C). For special effects, other temperatures can be used. Exact mixing procedures and close temperature control are needed to create consistent results. Firing, or "burning," is done in large kilns heated by gas or oil.

Brick production begins by mining clay or shale, and then crushing, pulverizing, screening, and mixing the material with water. The old "soft mud process" used considerable water, and the bricks were formed in a mold. In the more common "stiff mud" process, bricks are extruded from a die and cut with a wire. After drying, the bricks are fired, inspected, and stacked.

Bricks are available in a wide variety of colors and finishes. Most clays fire to either buff or red color. Red color results from the presence of iron oxide in the clay. Higher kiln temperatures cause darker colors and sometimes shape distortion. Some color variation is inevitable in brick production, due to differences in kiln temperature and chemical composition of the raw clay.

"Flashing" in a very hot kiln drives off the oxygen and forces the iron oxides to undergo the chemical process called reduction. The result is a purplish color on the bricks.

Textures are applied to the clay before firing by scoring, scraping, or impressing with rollers. Face brick can be glazed or coated to give desired colors.

Bricks are shipped and delivered in cubes of 200 to 500 units. The cubes which have slots that accept the tines of a forklift. Cubes are wrapped in steel banding and/or plastic for easy handling. Each cube is assembled of packages containing as many as 100 bricks. Packages may be wheeled around the site on special brick and block trucks.

ABSORBENCY

Some bricks are very absorbent and tend to draw excessive water from the mortar, reducing the time available for adjustment and impairing hydration. Test for excess absorption, also called suction, by sprinkling some water on the brick surface. If the water disappears in one minute, the brick should be dampened before use. A more accurate test is to mark a circle one inch in diameter with a wax crayon, using a quarter-dollar as a guide. Quickly squirt 20 drops of water from an eye dropper into the circle, allowing no water to escape the circle. If the water is entirely absorbed within 90 seconds, the brick can be considered high-suction and should be treated accordingly.

Bricks can be dampened by laying them out and sprinkling with a hose or lawn sprinkler. Another method is to dip bricks in a pail or trough. With either technique, experiment to find out how much soaking is required. After wetting, allow the water to sink in. If you use the brick when the surface is wet, mortar will ooze from the joint, the brick will float on the surface, and the bond will be weak.

GRADE

Bricks are graded according to their ability to resist weather. Suppliers generally carry only those grades suited to their area.

Brick grades	
Grade SW (severe weathering):	Suited to the most extreme exposure conditions, where freezing and thawing are expected. This grade is sold in moist, northern areas of the United States. It is most suitable to below-grade applications. Compressive strength is 3,000 psi.
Grade MW (moderate weathering):	Can withstand freezing in dry climates. Compressive strength is 2,500 psi.
Grade NW (no weathering):	Cannot be exposed to much weathering. Used in dry climates with some frost or in damp climates with no frost. Also used in interior work and backup walls. Compressive strength is 1,500 psi.

SIZE

Bricks are described by two sizes: actual size and nominal size. Actual, or manufactured, size can be measured directly from the unit. Nominal size is found by adding the dimension of one mortar joint to the actual size. The terminology for bricks is not entirely standardized, so it is best to check size as well as name when ordering.

As an example of the relation between actual and nominal size, in a wall with 3/8-inch joints, standard modular brick that is 2-1/4 inches high, 3-5/8 inches wide and 7-5/8 inches long would have nominal dimensions of 2-2/3 inches high, 3-5/8 inches wide and 8 inches long. See Table 4.3 for details on nominal and actual sizes of modular brick.

Modular bricks fit modules to simplify design and allow the use of standard windows, doors, and other building elements. Not all bricks conform to the modular convention. In recent years, a great number of larger bricks have been developed. These bricks increase efficiency on the job and look attractive in large buildings.

TERMINOLOGY

Masons have a special jargon to describe the different sizes, shapes, parts, and positions of bricks. These terms are easier to portray than to explain (see Fig. 4.5).

TYPES

Various brick types can be used for different functions. Not all types are sold in each size or color combination; check local suppliers.

BUILDING (COMMON)

Building brick is made from clay that has not been treated for uniform color or texture. Building brick is usually not fired as hard as face brick; it is often used for backing walls. The term "common brick" often refers to concrete brick.

CLINKER

Clinkers are produced by excessive heat in the kiln. They are rough, hard, and likely to be dark and deformed. Clinkers can be used for decorative effects.

FACE

Face brick has special colorings and surface textures. It is suited for weather exposure and is generally considered the most attractive brick. Three grades are available:

FBX: for walls requiring near-perfect units.

FBS: for general use on both interior and exterior work; some variation in color and surface permitted.

FBA: selected for pleasing variety of color and texture; less uniform than the other grades; used for architectural effects.

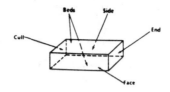

a)

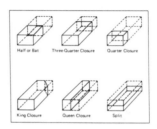

b)

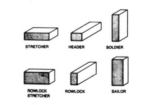

c)

Fig. 4.5   Brick Terminology: a) surfaces b) parts c) positions

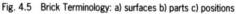

FIREBRICK

Firebrick, or refractory brick, is compounded from fire clay to resist high temperatures. Firebrick is used in fireplaces, boilers, kilns, and other high-temperature applications. Firebrick should be mortared with refractory cement, which is also designed to withstand high temperatures.

Firebrick is commonly sized at 9 inches × 4-1/2 inches × 2-1/2 inches; also available are soaps, 9 inches × 2-1/2 inches × 2-1/4 inches, and splits, 9 inches × 4-1/2 inches × 1-1/4 inches.

Courtesy of Brick Institute of America

TABLE 4.3
Sizes of Modular Brick[1]

Unit Designation	Nominal Dimension, in.			Joint Thickness in.	Manufactured Dimensions			Modular Coursing
	t	h	l		t	h	l	
Standard Modular	4	$2\frac{2}{3}$	8	$\frac{3}{8}$	$3\frac{5}{8}$	$2\frac{1}{4}$	$7\frac{5}{8}$	3C = 8
				$\frac{1}{2}$	$3\frac{1}{2}$	$2\frac{1}{4}$	$7\frac{1}{2}$	
Engineer	4	$3\frac{1}{5}$	8	$\frac{3}{8}$	$3\frac{5}{8}$	$2\frac{13}{16}$	$7\frac{5}{8}$	5C = 16
				$\frac{1}{2}$	$3\frac{1}{2}$	$2\frac{11}{16}$	$7\frac{1}{2}$	
Economy 8 or Jumbo Closure-	4	4	8	$\frac{3}{8}$	$3\frac{5}{8}$	$3\frac{5}{8}$	$7\frac{5}{8}$	1C = 4
				$\frac{1}{2}$	$3\frac{1}{2}$	$3\frac{1}{2}$	$7\frac{1}{2}$	
Double	4	$5\frac{1}{3}$	8	$\frac{3}{8}$	$3\frac{5}{8}$	$4\frac{15}{16}$	$7\frac{5}{8}$	3C = 16
				$\frac{1}{2}$	$3\frac{1}{2}$	$4\frac{13}{16}$	$7\frac{1}{2}$	
Roman	4	2	12	$\frac{3}{8}$	$3\frac{5}{8}$	$1\frac{5}{8}$	$11\frac{5}{8}$	2C = 4
				$\frac{1}{2}$	$3\frac{1}{2}$	$1\frac{1}{2}$	$11\frac{1}{2}$	
Norman	4	$2\frac{2}{3}$	12	$\frac{3}{8}$	$3\frac{5}{8}$	$2\frac{1}{4}$	$11\frac{5}{8}$	3C = 8
				$\frac{1}{2}$	$3\frac{1}{2}$	$2\frac{1}{4}$	$11\frac{1}{2}$	
Norwegian	4	$3\frac{1}{5}$	12	$\frac{3}{8}$	$3\frac{5}{8}$	$2\frac{13}{16}$	$11\frac{5}{8}$	5C = 16
				$\frac{1}{2}$	$3\frac{1}{2}$	$2\frac{11}{16}$	$11\frac{1}{2}$	
Economy 12 or Jumbo Utility	4	4	12	$\frac{3}{8}$	$3\frac{5}{8}$	$3\frac{5}{8}$	$11\frac{5}{8}$	1C = 4
				$\frac{1}{2}$	$3\frac{1}{2}$	$3\frac{1}{2}$	$11\frac{1}{2}$	
Triple	4	$5\frac{1}{3}$	12	$\frac{3}{8}$	$3\frac{5}{8}$	$4\frac{15}{16}$	$11\frac{5}{8}$	3C = 16
				$\frac{1}{2}$	$3\frac{1}{2}$	$4\frac{13}{16}$	$11\frac{1}{2}$	
SCR brick[2]	4	$2\frac{2}{3}$	12	$\frac{3}{8}$	$5\frac{5}{8}$	$2\frac{1}{4}$	$11\frac{5}{8}$	3C = 8
				$\frac{1}{2}$	$5\frac{1}{2}$	$2\frac{1}{4}$	$11\frac{1}{2}$	
6-in. Norwegian	4	$3\frac{1}{5}$	12	$\frac{3}{8}$	$5\frac{5}{8}$	$2\frac{13}{16}$	$11\frac{5}{8}$	5C = 16
				$\frac{1}{2}$	$5\frac{1}{2}$	$2\frac{11}{16}$	$11\frac{1}{2}$	
6-in. Jumbo	4	4	12	$\frac{3}{8}$	$5\frac{5}{8}$	$3\frac{5}{8}$	$11\frac{5}{8}$	1C = 4
				$\frac{1}{2}$	$5\frac{1}{2}$	$3\frac{1}{2}$	$11\frac{1}{2}$	
8-in. Jumbo	4	4	12	$\frac{3}{8}$	$7\frac{5}{8}$	$3\frac{5}{8}$	$11\frac{5}{8}$	1C = 4
				$\frac{1}{2}$	$7\frac{1}{2}$	$3\frac{1}{2}$	$11\frac{1}{2}$	

[1] AVAILABLE AS SOLID UNITS CONFORMING TO ASTM C 216- OR ASTM C 62-. OR, IN A NUMBER OF CASES, AS HOLLOW BRICK CONFORMING TO ASTM C 652-.

[2] REG. U.S. PAT. OFF., SCPI.

When building boilers, kilns, or digesters, masons use many special sizes, shapes, and varieties of firebrick and other refractory materials.

GLAZED

Glazed bricks are face bricks that were treated with liquid glazing before firing. The resultant glassy surface is shiny, hard, and nonporous. Glazed brick is used in laboratories, hospitals, kitchens, and bathrooms because it is easy to clean. It can also be used for its shiny, decorative effect.

PAVING

Paving brick is durable and solid, with no holes or cores. The material is used in driveways, walkways, patios, and hearths. Paving brick ranges in thickness from 2-1/4 inches to 5 inches and thicker.

Two systems are used to classify paving brick: by durability to wear, and by capacity to resist weather exposure. A mason must take both classifications into account when specifying paving brick.

Traffic	
Type 1:	High-traffic areas, such as entrances to public buildings and driveways
Type 2:	Intermediate traffic, such as floors
Type 3:	Residential areas, such as patios and hearths.

Weathering	
Class SX	Resists high moisture and freezing.
Class MS	Used in exterior areas not subject to freezing.
Class NX	Used for interior work; the brick is sealed or coated before use.

SOLID

During manufacture, cores are removed from most bricks to improve the brick's drying and firing characteristics and to reduce weight. Bricks with voids are "solid" as long as the voids cover no more than 25 percent of their area.

SPECIAL SHAPES

Decorative or functional shapes can be ordered from some brick manufacturers. Corners with angles other than 90°, arch bricks, window sills, and various types of coping are available from stock or custom-made. Some suppliers will furnish handmade bricks with an antique look.

THIN BRICK (VENEER)

Thin brick is a recent innovation that is used strictly for veneer. The face size usually matches those of conventional bricks, so a well-laid wall resembles a conventional wall. Modular brick (2-2/3 inches × 8 inches nominal size) and economy (50 percent longer and higher) are both available. Thin bricks are usually 1/2 inch to 1 inch thick. Special pieces are needed for corners and other locations where more than the face will be visible.

Thin brick may be applied with organic adhesive or epoxy in the "thin set" method, which is only applicable to interior applications. The "thick set" method uses a portland cement plaster over wire lath or rough masonry substrate. Prefabricated panels are also possible (see *Walls, veneer*, p. 125).

USED (SALVAGED)

Used brick can be an attractive and economical material, but it should not be used where weathering is a problem, as it can rarely be made watertight. Salvaged units with excessively high or low suction will produce a poorly bonded wall subject to moisture penetration. Attention should be paid to the presence of salmons and clinkers, which should be used with extreme care if at all. Old mortar must be removed thoroughly or it will interfere with the new mortar bond.

Building codes require that salvaged brick pass much the same tests as new brick. For example, the National Building Code, section 1401.2, states:

Second-hand units: Bricks and other second-hand masonry units which are to be reused, shall be approved as to quality, condition and compliance with the requirements for new masonry units. The unit shall be of whole, sound material, free from cracks and other defects that would interfere with its proper laying or use, and shall be cleaned from old mortar before reuse.

Some new bricks are made to resemble used brick, and they can be used in exterior applications.

CAULKING

Caulks are used to fill voids between adjacent building materials to increase weather resistance. Caulking materials must be able to expand and contract in response to building movements and remain tightly bonded to the structure. In damp climates and wet locations (such as near the ground) mildew resistance is especially important.

A good caulk must:

- *adhere tightly to the substrate.*
- *remain workable through the entire service temperature range.*
- *form a tough, flexible skin while the interior remains flexible.*
- *be able to stretch and compress with movements in adjacent building elements.*
- *have very low volatility.*
- *resist weathering, mildew, and other biological actions.*
- *be able to be tooled to an attractive appearance.*
- *take paint after setting (desirable but not always essential).*
- *be washable with water.*

CAULKING TYPES

Many grades and types of caulks are available, and more are being developed by company laboratories. Caulks are available in various colors. White is most common, but various browns and sometimes grays are available. Clear caulks need no painting because they allow the base materials to show through and are inconspicuous in service.

Latex caulks are extremely workable, low in odor and paintable. They are not as durable as silicon, although latex can be compounded with other materials, such as acrylic or vinyl, to improve durability. Latex can be cleaned with water and produces little odor while drying.

Silicon caulks are the most durable, with some having guarantees as long as fifty years. Most silicon caulks cannot be painted, although silicon compounded with acrylic can be.

Butyl rubber caulks are suited to below-grade applications.

Polyurethane foam sealants are sold for filling and insulating hard-to-reach areas. The foam expands for a while after application, so it is best to fill the void slightly less than half-full and allow the foam to expand. The foam may be cut with a knife and painted after setting.

Elastomeric sealants are suggested for joints with a great deal of movement, such as expansion joints. For joints with less movement, such as around jambs, use a solvent-based acrylic sealant or butyl caulk. Oil-based caulks are not suggested for use with masonry.

CHOOSING CAULKS

Consider the following factors when choosing a caulking:

- *application and service temperature*
- *moisture conditions*
- *cost*
- *ease of application*
- *compatibility with adjacent building materials*

COVERAGE

Most caulks are available in 10.3-fluid-ounce (.304-liter) tubes. One tube will cover approximately 50 feet of a 1/8-inch × 1/4-inch bead; this is roughly enough to caulk the exterior of five windows.

LINEAR COVERAGE OF CAULKING IN FEET

Joint depth in inches	Joint width in inches			
	1/16	1/8	1/4	1/2
1/16	423	211	105	52
1/8	211	105	52	26
1/4	105	52	26	13
3/8	70	35	17	9

COATINGS

A great many materials are used to increase the ability of masonry materials to resist water. The term "waterproofing" is often used but is misleading because most of these materials increase resistance to water but do not make the masonry waterproof. Membrane compounds, however, do waterproof as long as they remain in good condition and the structure remains sound and stable. Use care when applying moisture-resistant coatings because certain building elements must be able to breathe, and coatings can cause them to degrade.

The major source of above-grade moisture is wind-driven rain. The need for a moisture-resistant coating increases along with wind speed and the quantity of rain. Buildings near large bodies of water or in damp climates have a great deal of wind-driven rain.

Below-grade moisture depends on soil drainage and the amount of water present. The amount and type of below-grade moisture protection needed varies depending on the application.

The following are some common moisture-resistant materials employed in masonry.

BENTONITE

Bentonite is a very fine clay useful for waterproofing in below-grade locations where soil humidity will never be less than 50 percent. When in contact with water, bentonite swells up to form a gel that fills cracks and prevents passage of water. It cannot be used in areas exposed to air because it will dry out and lose its resistance to moisture.

Trowel and spray grades of bentonite are available. Coats should be about 3/16 inch thick. Because soil movement can damage the coating, a polyethylene sheet is sometimes laid over the bentonite as protection.

Bentonite is also available in cardboard-coated panels that are nailed or glued with roofing cement to the foundation. In time, soil bacteria degrade the cardboard, leaving a smooth coating of bentonite on the wall.

BITUMINOUS SEALERS

Bituminous sealer compounds are used to protect below-grade masonry from water. The compounds can be sprayed or trowelled onto the surface. Some will adhere to fresh concrete and damp surfaces. Follow manufacturer's instructions. Bituminous sealers are also used for laying the first course of glass block.

ELASTIC COMPOUNDS

Elastic sealers are used for sealing entire surfaces of concrete, concrete block and brick, especially surfaces with hairline cracks. Acrylic emulsion finishes are compounded to allow a certain amount of flexibility after setting. The material resists ultraviolet degradation and allows water vapor to pass through. Gaps must be filled with special knife-grade material before applying elastic sealant.

   Elastic sealant adheres to sound, clean masonry substrates. Temperature and humidity must be within certain ranges during application. Follow the manufacturer's directions exactly.

EXPANSION COMPOUNDS

Expansion compounds contain a metallic element, often cast iron. The compound is brushed into a surface and allowed to expand. The compounds have excellent adhesion and will clog cracks and pores. The compounds are often used below grade because they leave a rust-colored surface. However, they may be finished with suitable material if used above grade. Gas-producing grout and expansive cement work in a similar fashion. Follow manufacturer's instructions carefully. In some cases, the manufacturer will send a representative to check that conditions and application techniques are appropriate.

HYDRAULIC CEMENT

Hydraulic cement is a compound that expands as it sets. The material sets under water and expands to fill a hole with water running through it, so it can be used to waterproof cracks in below-grade structures. Hydraulic cement can repair blisters, honeycombs, and other construction faults in poured concrete. It is also used to set posts, bolts, railings, and other metal elements in concrete, and to seal around pipes and other fittings protruding from masonry and concrete.

The material is sold dry and mixed with water immediately before use. One pound will fill about 15 cubic inches of holes (230 cm³ or a crack 3/4 inches × 3/4 inches × 28 inches).

SILICON SEALER

Several silicon sealers are available as water-repellent treatments for masonry. Properly applied, silicon compounds:

- *resist moisture penetration, reducing damage to walls and building interiors.*

- *allow moisture to breathe from a building.*

- *minimize efflorescence, spalling, and cracking.*

- *keep surfaces clean.*

Silicon sealants are intended to be applied to above-grade material with moderate porosity. Masonry should cure about thirty days before application. Silicon will not bridge gaps, so they should be repaired first. Surfaces should be free of loose material, stains, and efflorescence. Extremely porous materials, such as cinder block, may need a grout coat before treating.

Handle silicon with care. Some compounds are flammable. Work with adequate ventilation or use a respirator. Test an inconspicuous panel first to ensure against discoloration.

Use a low-pressure spray or brush for application. When spraying, flood the surface to ensure complete coverage. Some materials require two treatments. Do not apply after heavy rains. Observe the manufacturer's recommendations for proper temperature during application.

Coverage per coat ranges from 50 square feet per gallon for very porous substrates to 200 square feet per gallon for smooth, dense substrates. Silicon is invisible after application and lasts for five to ten years.

CHOOSING A COATING

Various coatings are designed for the different applications—above- and below-grade, and interior and exterior work. Some are designed to be neutral in appearance or to improve the appearance, while others are used where appearance is not important.

Coatings may do more harm than good. The Brick Institute of America suggests these dangers in coating walls:

- *The treatment may not prevent further water entry because of cracks or poor joints.*
- *The treatment can cause or increase disintegration of bricks by causing the buildup of salt crystals inside the units.*
- *The coating may prevent the removal of stains and efflorescence.*
- *The coating may cause damaging efflorescence. Treating an efflorescing wall with silicon or acrylic coatings may cause soluble salts to build up near the surface, which can cause spalling. The source of moisture must be controlled or such an application will cause more harm than good.*
- *Coating may prevent future tuckpointing if needed.*

These considerations should influence the choice of a coating:

- *Longevity—especially when the coating will be covered by earth or another barrier.*
- *Wear resistance—on floors carrying traffic.*
- *Ability to form a continuous film under job conditions and to shift with the structure if needed.*
- *Resistance to ultraviolet light in sunlight and ozone in the air.*
- *Ability to bond with the substrate.*
- *Appearance of the finished coat.*
- *Compatibility with other building components, such as flashings and caulks.*
- *Weather conditions needed during application.*
- *Cost per square foot.*

PAINT

Paint is used to change the appearance of masonry walls and sometimes to increase resistance to weathering. Do not use waterproof paints on exterior walls because these walls must usually breathe to prevent moisture buildup. The type of paint chosen will depend on wall material and condition, weather, and conditions during application. The following are some considerations to evaluate before choosing a paint:

- *Compatibility with alkalinity in units and mortar. An alkali-resistant primer may be used to reduce damage caused by such alkalinity.*
- *Efflorescence must be controlled or it will damage the paint. Remove any efflorescence first, and wait to see if it recurs before painting (see* Troubleshooting, Efflorescence, *p. 295).*
- *Water on the surface will interfere with the application of many paints. In addition, persistent moisture will cause numerous problems with the paint and the wall.*

TYPES OF PAINT

Paint can be divided into cement-, solvent-, and water-based.

Cement-based paints contain portland cement. They are durable, strong-bonding, relatively difficult to apply, and useful in sealing porous structures. Portland cement-based paints are breathable yet they reduce the rate of water infiltration. Dark colors are more likely to show streaks than light colors.

Solvent-based paints must be applied over completely dry, clean surfaces. The major groups of these paints are oil-based, alkyd, synthetic rubber, chlorinated rubber, and epoxy. In general, solvent-based paints have low permeability to water and may be unsuitable for exterior application. This is especially true of oil-based and alkyd paints. Use solvent-based paints with care—the fumes are dangerous to your health and may be explosive.

Water-based (latex) paints are convenient to apply and usually breathable. Application with brush is best because it fills crevices most effectively. Latex paints may be applied in damp conditions, are alkali-resistant (in general), and can be cleaned up with water. Use caution over old paint that has chalked—either remove the chalk or use a paint that can wet the chalk.

CONCRETE

Concrete is a mixture of portland cement, stone, sand, and water. Portland cement was patented in England in 1824 and was named for its color, which resembled a stone found on the Island of Portland. Concrete contracts with a loss of moisture and expands with an increase of moisture. Shrinkage is negligible in very damp climates and severe in arid climates. Lightweight aggregates tend to shrink more than standard sand-and-gravel aggregates.

Concrete is an economical building material for bridges, beams, highways, foundations, floors, walls, masonry units, and many other uses. Concrete has great compressive strength, and its relatively low tensile strength can be compensated by embedding steel reinforcement in the concrete.

Masons use concrete for footings, bond beams, and vertical reinforcing of block walls, and for slabs under fireplaces, cast-in-place beams and lintels, sills and coping. Concrete block is made of concrete with a relatively fine coarse aggregate (see *Portland cement*, p. 80).

COPING

Coping is used to cap parapet and other walls with exposed tops. Coping is usually about 1 inch wider than the wall on each side, and has a "drip" shape in the outside bottom edge. The purpose is to cause the water to drip directly from the edge, and not run down the wall. In general, the number of joints in coping should be minimized to reduce the chance of water penetration.

Coping may be made of stone, concrete, brick, terra cotta, or metal. Stone coping can be cut to fit. Concrete coping can be cast in place or precast. Cast-in-place concrete copings must have provisions for shrinkage during drying. Sections of precast copings are cast in 4 feet to 8 feet long modules, depending on the weight of the particular cross-section. Fired clay copings are cast in 24-inch sections. Ends, 90° corners, and tee formations are available for fired clay coping. Metal coping has a much greater coefficient of expansion and must have provision for movement.

Care must be taken to provide flashing beneath the joints of most coping to ensure a watertight wall.

FLASHING

Flashing is used to seal joints between masonry and other material. Flashing is specified in blueprints. Typical locations include the joint between a chimney and roof, sills, lintels, the bottom of veneer walls, the intersection of parapet walls and roofs, and other projections or recesses in walls. On some jobs, flashing is fabricated by sheet metal trades. Some types of flashing may be purchased prefabricated.

Many materials can be used for flashing masonry. Aluminum and lead are subject to degradation from alkalies and should not be used. Felt impregnated with asphalt may be damaged during construction and becomes brittle with age, so it should not be used. The following materials are used for flashing masonry:

1. Galvanized steel is commonly used but it may corrode in contact with fresh mortar, especially if it is bent. Minimum thickness should be 0.015 inch.

2. Plastics, including vinyl, are very common flashing materials. This is a wide group, and performance varies with the specific material. Plastic must resist ultraviolet light and be compatible with the alkalinity in mortar and any joint sealants used.

3. Copper has the best durability but is incompatible with mortars containing chlorides. The weight of copper sheet metal should range from 10 to 20 ounces. Copper may stain adjacent masonry but is not affected by alkalies in mortar.

4. Composite flashings are constructed to take advantage of different materials. Copper is sometimes bonded to glass fibers for a workable, durable material. Copper can also be laminated with polyethylene, asphalt, or kraft paper. Consult the manufacturer for details on applications of these flashings.

Flashing is sealed with caulkings made of silicone, urethane, and butyl. Asphalt-based roof cement is frequently used. Galvanized steel and copper can be soldered to themselves for a permanent joint.

FLUE LINER

Chimney flues are usually lined with fired clay liner to control damage and danger from chimney fires and moisture. Flue liner, also called flue tile or flue, is formed in rectangular or circular cross-sections. The liner is designed to resist fires of 1800°F (980°C), which can occur during chimney fires in chimneys serving wood-burning stoves and fireplaces.

Flue liners also resist moisture damage, an important point since water vapor forms when fossil fuels are burned. Water vapor can condense and freeze under some conditions. Vitreous clay tile is best for flues due to its great resistance to moisture and ease of cleaning.

Flue tiles must be at least 5/8 inch thick. Codes may require that sections of flue be bonded with refractory cement, not regular mortar. Joints must fit tightly. An air gap is needed between the liner and the chimney walls to allow for expansion. If a class A chimney is required, a flue liner is practically always needed. Check local codes. A 2-inch air space is required by many building codes between the chimney and all flammable building material.

Most sections of round and rectangular flue are 24 inches long. Rectangular flue sizes range from modular 4 inches × 8 inches to 20 inches × 24 inches. In round flue, the nominal inside diameter ranges from 6 inches to 36 inches.

TABLE 4.4
Flue Liner Dimensions

Nominal Size in.	Inside Perimeter, P_1 ft.	Equivalent Diameter, D in.	Minimum Area, A_F sq. ft.
8×12	2.7	9	0.42
12×12	3.0	10	0.56
12×16	3.6	12	0.78
16×16	4.2	14	1.08
16×20	4.8	16	1.38
20×20	5.3	18	1.78
20×24	5.9	20	2.16
24×24	6.5	22	2.64

Flue sizing can be confusing. A modular 8-inch × 8-inch flue is actually 7-1/2 inches × 7-1/2 inches, so it can fit inside an 8-inch × 8-inch (actual) inside dimension chimney. However, a standard 8-inch × 8-inch flue may be closer to 9 inches square, and an 8-inch × 12-inch flue is likely to be closer to 9 inches × 13 inches.

INSULATION

Most masonry material has little long-term resistance to the flow of heat, although it does reduce swings in temperature. With the increasing consciousness about energy conservation, both architects and building codes call for additional insulation. In recent years, many insulation systems have been developed to speed construction and meet strict energy conservation requirements. Consult the manufacturer or distributor for more information about insulation systems available in your area.

Insulation is graded by R-factors—the higher the R-factor, the greater the resistance to heat transfer. R-factors may be added together to find the total resistance to heat flow.

POURABLE

Several materials, including vermiculite, perlite, and styrofoam granules, may be poured into the cavities of concrete blocks or the collar joint of a cavity wall. Note that insulation must be poured into cavities before bond beams are cast and window sills are placed. Continuous beaded styrene liner pieces are available for insertion into block cores.

Use only insulation materials that cannot absorb water. A good pourable insulation is nonsettling, water-repellent, and rot-proof, and must be able to pour freely so it can entirely fill a cavity. The Brick Institute of America suggests these two pourable insulations: water-repellent vermiculite, produced and marketed by members of the Vermiculite Association, and silicone-treated perlite loose fill insulation, produced and marketed by members of the Perlite Institute, Inc.

APPROXIMATE CUBIC FEET OF POURABLE INSULATION NEEDED TO FILL 100 SQUARE FEET OF WALL	
6-inch block	17.2
8-inch block	27.2
10-inch block	37.6
12-inch block	48.4
1-inch cavity	8.4
2-inch cavity	16.8
2-1/2-inch cavity	20

RIGID

Masonry walls may be insulated with rigid boards of polyurethane or polystyrene. These boards offer great resistance to heat flow and can be installed quickly and at low cost. Unlike pourable insulation, board insulation prevents heat transfer through block webs, so it is more effective at reducing heat transfer from an entire surface.

Boards are available in widths of 16 inches, 24 inches, and 48 inches, commonly in sheets 8 feet long. Thickness ranges from 1/2 inch to 2 inches. Tongue and groove sheets simplify application and reduce air infiltration at joints.

Styrofoam panels are used to insulate foundation exteriors. Boards can be fastened with adhesive. Mechanical fasteners are recommended for below-grade installations (if the insulation does not rest on a sill) or any place a brush-on coating is used. To prevent sunlight degradation and mechanical damage, coat the insulation between the siding and grade. Brush-on and trowel-on coatings are used, either directly on the board or over wire lath.

Board insulation can also be inserted in cavity walls. Use care to place ties so they will not interfere with the insulation. You may need to use 16-inch or 24-inch-wide boards to leave room for wall ties.

EXTERIOR SYSTEMS

Exterior insulation finishing systems (EIFS on blueprints) combine foam insulation boards with mesh and cementitious material. These systems are becoming more popular due to their effectiveness, cost, and ease of application. Methods and components vary with the system and building project.

Insulation is first bonded to the wall with fasteners and/or adhesive, then a mesh is applied over the insulation. Finally, a synthetic finish of the desired color is trowelled on and textured as specified. The system can be retrofitted to existing buildings and used on new construction. Panels can be constructed on the ground or inside a building, allowing construction to proceed during bad weather.

MOVEMENT JOINT

Two types of joints are used to allow differential movement in masonry structures. 1) Control joints are used to prevent cracking in concrete structures due to shrinkage drying and subsequent expansion from various sources. 2) Expansion joints are used to allow brick structures to expand as they gain moisture (clay products are dry when laid and shrinkage drying is not a problem). Architects are responsible for specifying the location of movement joints.

CONTROL JOINT

Control joints permit contraction from shrinkage drying and subsequent movement of the structure. An effective control joint must 1) allow movement between the adjacent building elements, 2) provide structural integrity, and 3) be weatherproof.

Either standard or sash blocks can be used with special caulking and backup materials to form a control joint. Control joints can be made of soft materials such as neoprene or rubber, or of hard materials such as copper or plastic. Care must be taken to ensure that no mortar gets inside the joint as this can prevent movement. Caulking is frequently inserted into the joint after the wall is laid. Caulking rope or backer rod made of closed-cell foam is frequently used to back up caulking in a deep joint.

Control joints are located at places where cracks would normally appear, such as on long walls, near windows, and near where walls change thickness. Control joints in backing wythes should be paired with movement joints in facing wythes to allow the walls to move independently of each other (see Fig. 4.6).

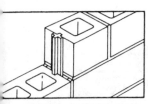

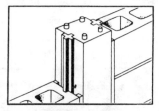

Gasket Used With Standard Sash Block

Gasket Used With Concrete Column or
Unbonded Pilaster

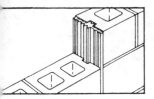

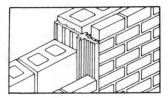

Wide Flange Gasket for 6" and 8" Walls
(Standard Sash Block)

Wide Flange Gasket Used in Composite Wall
(With Concrete Brick)

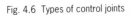

Fig. 4.6 Types of control joints

Courtesy of DUR-O-WAL, Inc.

EXPANSION JOINT

Expansion joints permit differential movement in clay masonry walls due to thermal expansion. Expansion joints may be made of copper, neoprene, extruded plastic, or premolded foam rubber or plastic. The joint must 1) prevent water entry, 2) provide a structural break, and 3) be elastic enough to compress and expand according to building movements. Expansion joints are formed by leaving a vertical joint empty of mortar and filling it with the premolded joint material (see Fig. 4.7).

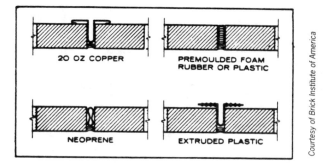

Courtesy of Brick Institute of America

Fig. 4.7 Types of expansion joints

MORTAR

Mortar is made of portland cement, lime, sand, and water. Masonry cement, a mixture of portland cement and lime containing various additives to increase workability, can be substituted for the portland cement and lime in the mixture.

Each element of the mortar has a role in creating a strong mortar. Sand provides a framework that is bound into place by the cement and lime. Portland cement hydrates in combination with water to bind the sand, and provide early strength and compressive strength. Lime increases workability, elasticity, and water retentivity and hardens in the presence of air, filling in hairline cracks over time. Water provides workability and is needed for hydration of the portland cement.

DESIRED PROPERTIES

Mortar has a series of desired properties. Some vary according to the building assignment, others are needed to make the mortar easy to work.

WORKABILITY

Mortar that is easy to place where it is needed is "workable." Experienced masons know workable mortar when they feel it. This mortar flows when spread; adheres to the trowel and vertical surfaces during placement; and retains water during placement and setting.

Mortar must set neither too quickly nor too slowly. This "hardening rate" depends on the weather, the nature of the masonry materials, and the chemicals present in the cement. A mortar that hardens too quickly is unworkable because the units cannot be placed, plumbed, and aligned before the mortar hardens. Mortar that remains plastic too long will 1) not support the weight of a wall under construction, 2) allow the units to "swim" instead of remaining in the proper location and 3) in cold conditions, freeze before setting.

APPEARANCE

Pigments are frequently added to mortar to produce colors other than the standard grey and optional white. In any case, mortar must be of uniform color to achieve good visual impact. Several factors help ensure color uniformity: using consistent aggregate and cement throughout the job, measuring ingredients carefully and mixing them thoroughly, and tooling all joints when they have the same degree of set.

DURABILITY

Mortar must be durable so it will remain bonded to the units and resist mechanical stresses on them. Masonry joints often must resist weathering and freeze–thaw cycles. In locations that are saturated with water, such as in wet soil, under downspouts, on horizontal ledges, or on walls with little overhang, freeze–thaw cycles are especially severe.

Because the various grades of mortar have differing durability, mortar should always be chosen to match the situation. Improper mortar and materials, and unsound building practices will decrease durability. Choosing correct materials and using them properly will meet most building situations.

STRENGTH

Increasing the cement content will increase the compressive strength of mortar. Compressive strength decreases with the presence of excess mixing water, entrained air, and lime. In general, stronger mortar is less workable, or more expensive and prone to shrinkage. Mortar should be chosen as a compromise between strength, economy, shrink-resistance, and workability (see Table 4.5).

The Brick Institute of America recommends the following "basic rule" for selecting mortar: "Never use a mortar that is stronger (in

TABLE 4.5

Assumed Compressive Strength of Brick Masonry

Compressive Strength of Units, psi	Assumed Compressive Strength of Brick Masonry f'_m psi					
	Without Inspection			With Inspection		
	Type N Mortar	Type S Mortar	Type M Mortar	Type N Mortar	Type S Mortar	Type M Mortar
14,000 plus	2140	2600	3070	3200	3900	4600
12,000	1870	2270	2670	2800	3400	4000
10,000	1600	1930	2270	2400	2900	3400
8,000	1340	1600	1870	2000	2400	2800
6,000	1070	1270	1470	1600	1900	2200
4,000	800	930	1070	1200	1400	1600
2,000	530	600	670	800	900	1000

COURTESY OF THE BRICK INSTITUTE OF AMERICA

compression) than is required by the structural requirements of the project. Always select the weakest (in compression) mortar that is consistent with the performance requirements of the project." However, because portland cement imparts a great deal of resistance to weathering, there are reasons for using a stronger mix than structural needs alone would dictate.

Air content above 12 percent has been found to reduce tensile bond strength. For this reason, air-entraining additives or portland cements of the air-entraining type are not recommended for unit masonry.

INGREDIENTS

ADMIXTURES

In some cases, it is desired to change the properties of mortar to suit special situations. Admixtures are used to achieve a faster or slower set, to make the mortar more water-resistant, and to increase workability (see *Weather, Cold,* p. 297).

Accelerators. Calcium chloride is often used to accelerate the set of concrete and sometimes to accelerate mortar. Calcium chloride can cause efflorescence and shrinkage, and it will corrode metal reinforcement and ties. If used, the weight of calcium chloride should not exceed 2 percent of the weight of cement. Use it only in mortar that will not be in contact with steel. (See *Weather, Cold.*)

Noncorrosive accelerating agents, containing no calcium chloride, are available. These products speed up hydration and increase early strength. In the process, they liberate heat from the chemical reactions within the mortar, and this helps protect the mortar from freezing. The rate of addition generally depends on the ambient temperature at the job site. Consult manufacturers for more information.

Water-reducing agents increase workability at low temperatures. They also increase set mortar's resistance to destruction from freeze–thaw cycles.

Plasticizers. Some admixes increase the plasticity of mortar while reducing the need for mixing water and speeding up the initial set. These qualities are helpful during cold weather, because reduced water content will lower the danger of freezing of fresh mortar.

Bonding agents. Bonding agents are used to increase adhesion of surface coatings and to reduce cracking in stucco. The bonding agent may replace part of the mixing water or be mixed with mixing water prior to mixing it with aggregate and cement. Consult the manufacturer's instructions.

Retardants. Retarding additives are used to slow the setting of mortar. These are useful for hot temperatures, where the fast set of mortar makes it difficult to position units correctly. Some additives delay the set for as long as 48 hours, allowing mortar to be batch-mixed and delivered to the site in ready-mix trucks. This permits accurate batching, reduces job-site labor and space requirements, and allows extra mortar from one day to be saved for use the following day.

Waterproofing. In recent years, a liquid polymeric waterproofing material has been developed for mortar and sold under several brand names. One quart of this material is added to two bags of Type M or S masonry cement or to one bag each of portland cement and lime. Normal joint tooling is used, except that raked joints are not permitted. The use of other admixtures is discouraged with the Dry-Block System. Weep holes are required above all lintels and other obstructions. If this waterproofing material is used when laying waterproof masonry units, there is no need for further sealing. Follow the manufacturer's instructions when using this material (see *Block, Concrete*, p. 38).

COLORING

Colored mortar is used to either highlight the mortar or to make it blend with the masonry units. Masonry cement is available precolored or pigments may be added to standard masonry cement. The maximum amount of pigment permitted is 10 to 15 percent of the weight of the cement. When using carbon black, limit the pigment to 2 to 3 percent of the weight of the cement. Excessive pigment reduces mortar strength.

Precolored cements are good for large jobs because they simplify the task of ensuring consistent color. Mixing colored cement is very similar to mixing standard cement. White cement is made by using white mortar and, if needed, white silica sand (see *Masonry Cement, white cement*, p. 74). Check local availability for colors.

The following chemicals are widely used as pigments:

Desired color	Pigment
Red, yellow, brown, black	Iron oxide
Green	Chromium oxide
Blue	Cobalt oxide
Black, grey	Carbon

Pigments should be premixed with water. After mixing the batch, test whether the pigment is fully mixed. Flatten some of the mix under a trowel. Any streaks indicate inadequate mixing, and the batch should be remixed immediately. Avoid retempering a batch after it begins to set because this will tend to fade the color.

LIME

Although portland cement and sand would make a mortar with great compressive strength, it would not be very workable. Thus masonry cement combines hydrated lime with portland cement for strength and workability. (Hydrated lime does decrease the compressive and tensile strength of mortar.)

Hydrated lime is produced by heating crushed limestone, which leaves calcium oxide (CaO). When CaO is treated with water, hydrated lime (Ca(OH)) results. Lime is sold in 1-cubic-foot bags.

MASONRY CEMENT

Masonry cement, a blend of portland cement and other ingredients, is sold in bags containing 1 cubic foot (see *Portland Cement*, p. 80). Because different types of masonry cement have different densities, each type weighs a different amount per cubic foot.

Two specifications are used to describe the various types of mortar: proportion and property. The more common technique is proportion, a set of recipes for mortar composition. Property specifications are dependent on tests that show that a certain mortar has properties that enable it to meet the standards for the type in question. Thus, a Type M mortar, according to proportion specification, contains 1 part portland cement, 1/4 part lime, and 3-3/4 parts sand. Under a property specification, a Type M mortar must test out with the same results as Type M mortar prepared under the proportion specification. But the mortar might have different components than the proportion specification mortar.

Type M High-strength material for below-grade use, as in foundations and retaining walls. Contains 1 part portland cement, 1/4 part lime, and 3-3/4 parts sand. Weighs 80 pounds per cubic foot.

Type S Combines the highest tensile bond with good compressive strength. Recommended for above-grade, exterior use in reinforced and unreinforced masonry. Can be used between facing and backing. The high tensile bond allows use where strong lateral forces affect a wall. Also suitable where mortar adhesion is the only bond between facing and backing. Contains 1 part portland cement, 1/2 part lime, and 4-1/2 parts sand. Weighs 75 pounds per cubic foot.

Type N A middle-grade mortar for above-grade work where loads are not excessive. Especially useful for chimneys, parapet walls, and exterior walls with severe weather exposure. Contains 1 part portland cement, 1 part lime, and 6 parts sand. Has adequate compressive strength and extreme workability. Weighs 70 pounds per cubic foot.

Type O A low-strength mortar, recommended for non-loadbearing, interior walls. Not recommended for reinforced structures. Maximum compression should not exceed 100 pounds per square inch. Does not resist freezing and thawing and has little moisture resistance. Contains 1 part portland cement, 2 parts lime, and 3 parts sand. Weighs 70 pounds per cubic foot.

White cement. White masonry cement is recommended for setting light-colored stone because it does not stain. "Staining"—the migration of some metallic compounds from the joint to surrounding stones—can discolor stonework. If a very white mortar is desired, use white silica sand as aggregate because elements in other sands may discolor white cement. Test the proposed sand to be sure it will perform as expected.

TABLE 4.6
Selecting Masonry Mortars

Location	Building Segment	Mortar Type	
		Recommended	Alternative
Exterior, above grade	load-bearing wall	N	S or M
	non-load bearing wall	O	N or S
	parapet wall	N	S
Exterior, at or below grade	foundation wall, retaining wall, manholes, sewers, pavements, walks, and patios	S	M or N
Interior	load-bearing wall	N	S or M
	non-bearing partitions	O	N

SOURCE: ASTM DESIGNATION: C 270-86B
COURTESY OF AMERICAN SOCIETY FOR TESTING AND MATERIALS

SAND

Sand must be washed clean and free of silt, salt, and organic matter or the mortar will not be strong and durable. To test cleanliness, place a couple of inches of sand in a quart jar, fill with water, and shake. Wait an hour. If a scum appears on top of the sand, it has excessive silt and should not be used.

Both natural and manufactured sand are acceptable. Natural sand is dug from pits. Manufactured sand is ground from stone, gravel, or blast furnace slag. Natural sand tends to have rounded corners while manufactured sand has sharp corners.

Sand should have a combination of large and small particles for maximum workability and strength. Mortar composed of sand with only large particles will be difficult to work and require large amounts of cement; sand with too many small particles will make a weak and thin mortar.

Sand usually contains a certain amount of moisture—often 4 to 8 percent . This causes no problem unless the sand is nearly saturated, in which case the sand will make a slushy mortar even if no water is added in mixing. If rain is a possibility, protect the sand pile with plastic. During hot weather, cover sand to retain moisture.

Sand expands when damp—fine sand with 5 percent moisture has 40 percent greater volume than dry, fine sand. Wet sand shrinks and occupies much less volume than damp sand. If the moisture content of the sand changes significantly during a job, the mix proportions will change as well; another reason to cover sand against evaporation and rain.

TABLE 4.7
Recommended Sand Gradation Limits

Sieve Size	Percent Passing	
	Natural Sand	Manufactured Sand
No. 4 (4.75-mm)	100	100
No. 8 (2.36-mm)	95 to 100	95 to 100
No. 16 (1.18-mm)	70 to 100	70 to 100
No. 30 (600-μm)	40 to 75	40 to 75
No. 50 (300-μm)	10 to 35	20 to 40
No. 100 (150-μm)	2 to 15	10 to 25
No. 200 (75-μm)	0 to 2	0 to 10

SOURCE: ASTM DESIGNATION: C 144–84.
COURTESY OF AMERICAN SOCIETY FOR TESTING AND MATERIALS

WATER

Only clean water should be used for mixing mortar. In general, if water is fit to drink, it is suitable. For mixes that specify water content by pounds, figure 8-1/3 pounds per gallon.

To achieve maximum compressive strength in concrete, the water content should be kept to a minimum. Although this is not true of mortar, many masons incorrectly assume that mortar should also be mixed with the minimum amount of water. Actually, the Brick Institute of America recommends that mortar be mixed with the maximum amount of water consistent with the desired workability, because this will promote the maximum tensile bond between units.

Retempering is permitted to remove water lost to evaporation, as long as not more than two hours has passed since the original mixing. While retempering may reduce compressive strength somewhat, failing to retemper will greatly cut tensile strength.

SPECIAL-PURPOSE CEMENTS AND MORTARS

A variety of mortars with special properties are occasionally used by masons. Check with local building codes before using (see *Coatings, Hydraulic Cement*, p. 57).

EPOXY

A very strong bonding material can be made by combining an epoxy resin with a catalyst at the job site. Epoxy mortar is especially useful for bonding veneer material to backing walls.

GROUT

Grout is used to bond around reinforcements, especially in bond beams, collar joints, and inside the cells of masonry units. Reinforcing is always placed before the grout is added. Aggregate must be adjusted to the application. Aggregate size increases in larger cavities—the maximum size for most uses is 3/8 inch. In spaces more than 5 inches to 6 inches wide, use conventional concrete, as long as the aggregate is smaller than 1 inch.

Fine grout—used in grout spaces less than 2 inches wide—uses 1 part portland cement and 2-1/4 to 3 parts sand. A common recipe for coarse grout is 1 part portland cement, 2 parts sand, and 2 parts coarse aggregate.

Add enough water to produce a soupy grout that can be easily poured or pumped into place. Use puddling or vibration tools to ensure a complete fill (see *Grouting*, p. 240).

REFRACTORY

Refractory mortar, or fireclay, is used to lay firebrick inside combustion chambers and chimneys. The material is sold ready-mixed or dry. Firebricks are usually dipped in refractory cement, making a joint that is thinner than the average mortar joint.

SURFACE BONDING CEMENT

Surface bonding cement combines portland cement and glass fibers. It is used to bond concrete blocks and is applied to the wall after the units are stacked. Walls built with surface bonding cement have greater flexural strength and require less skill to build than standard mortared walls. Because no mortar joints are used, blocks do not conform to nominal size.

The surface bonding cement can be used as the final surface. Great resistance to water is claimed for these products, thus eliminating the need for surface treatments.

Surface bonding cement can also be used as a crackproofing coat over block, brick, and certain insulation materials. A one- or two-coat application is recommended. The material can be sprayed or trowelled onto the surface to coat both sides approximately 1/8 inch thick. One 50-pound bag of powder covers about 35–100 square feet of wall, depending on porosity, texture of the wall, finish, application method, and workmanship.

STONE MORTAR

Depending on its color and texture, stonework often looks better when laid with white or colored mortar instead of gray. Take care that the mortar does not contain excess water; many types of building stone have a low absorption rate and will not set properly in wet mortar.

TUCKPOINTING

Because shrinkage can be a problem when tuckpointing, mortar should be premixed, or prehydrated, to reduce shrinkage. The mortar type and color should be selected to match the color of the original mortar, the climate, and the type of brick. Two methods can be used: 1) Match the original color by allowing a sample of fresh mortar to dry on a trowel blade in the sun. Compare this color to the mortar in the wall under repair and adjust the pigments if necessary. 2) Mix a sample of mortar several days ahead of time. After the mortar has had a chance to set, break it and compare the broken surface to a freshly broken piece of the original mortar.

Prehydrate mortar by mixing sand and masonry cement dry in the mixing box. Add just enough water to make a dough-like ball and mix. Wait one or two hours and add enough water to make the mortar workable. Tuckpointing mortar will be drier than normal mortar while being worked and will shrink less during setting.

PAVING MATERIAL

A complete paving assembly consists of a base material, a cushion material, a setting bed, and a paving material. Many variations are possible within this basic setup.

Solid bricks and concrete patio blocks may be used as pavers over concrete or sand. Concrete pavers are sometimes pigmented to resemble brick pavers. Size and shapes vary with locality. A common size for dry use is 4 inches × 8 inches. Paving is laid either "dry"—without mortar, or "wet"—with mortar (see *Paving,* p. 262).

The three design factors for choosing pavers are 1) weathering, 2) traffic load, and 3) appearance. The most important characteristic for exterior paving is resistance to weathering. Dense, hard-burned brick is generally resistant to abrasion. Areas with vehicular traffic require a more abrasion-resistant brick than areas subject to pedestrian traffic only.

Rigid paving is laid over a concrete or asphalt base. Flexible paving is laid over compacted sand or gravel, or a mixture of sand and cement tamped into place (see Fig. 4.8).

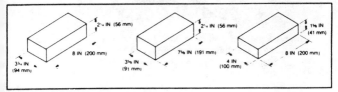

Fig. 4.8 Popular brick pavers

Courtesy of Brick Institute of America

BASE MATERIALS

Gravel is a common base which provides both support and drainage. Washed gravel is best because it prevents water movement. Unwashed gravel may contain fine particles that allow water to move through capillary action, causing efflorescence in the pavers. Unwashed gravel may also clog up over time and prevent further drainage.

Concrete bases are also common, and are suitable as long as proper practices are followed while forming and pouring the slab.

Asphalt bases are acceptable, but the mason should understand that little bond will occur between the asphalt and the mortar. Mortar laid on hot asphalt may flash-set, so work in hot weather should be avoided. Wood bases are unstable and prohibited by some building codes.

CUSHION MATERIALS AND SETTING BEDS

Various materials may be placed between the base and pavers to level the surface. About 1 inch to 2 inches of material is needed. Sand, 15- to 30-pound roofing felt, and a mixture of sand and portland cement are commonly used.

The setting bed is used in rigid paving to bond the pavers. Mortar, either standard or with additives for increased bonding strength, may be used. Type M mortar is recommended for exterior applications, Type S or N for interior use. Bituminous setting beds contain asphaltic cement and aggregate. This material is premixed and delivered hot to the job site.

PORTLAND CEMENT

Portland cement, a basic ingredient in concrete and mortar, is made by burning a combination of limestone and clay. The resulting clumps of material are ground into an extremely fine powder, then mixed with 2- or 3-percent gypsum to control the setting rate.

Portland cement is generally sold in 94-pound bags containing one cubic foot.

Five grades of portland cement are available:

Type I   General-purpose cement.

Type II   Has moderate resistance to sulfates that may be present in ground water. Creates less heat of hydration than Type I, a useful property in massive structures.

Type III   High early strength cement. Used in cold weather, where forms must be removed rapidly, or where the structure must be placed in service soon after pouring.

Type IV   Creates little heat of hydration, which is desirable in massive structures. Cures slowly.

Type V   Used in locations with a great deal of sulfate in soil or water. Gains strength more slowly than Type I.

Concrete made with air entraining portland cement contains millions of tiny bubbles of air. The air increases resistance to freeze–thaw cycles and to chemicals used to melt snow and ice. Portland cements of Types I, II, and III that contain air entraining additives are designated Types IA, IIA, and IIIA. However, because air entrainment reduces mortar bond to masonry units and reinforcement, do not combine these cements with lime and sand to make mortar.

REINFORCING AND TIES

Because masonry has great compressive strength, but little tensile strength, steel components are used to increase tensile strength for many applications, such as lintels, reinforcing, and other uses (see *Anchors*, p. 33).

ANGLE IRON

Angle iron is used to span openings. A single iron will support a veneer, while doubled angles will support two-wythe walls. The angle must bear at least 4 inches on the supporting wall at each end and the supported wythe must sit solidly on the angle, not overhang it. Building codes generally permit steel angles to be used for lintels as long as 8 feet. Longer lintels often must be fire-protected. This can be accomplished with a steel beam holding suspended or attached plates that support a masonry covering (see Fig. 4.9).

Angle iron is designated by an "L" symbol on blueprints. Size is described by the width of the legs of the angle and by the thickness of the steel. "L 4 × 4 × 1/4" describes an angle iron with two 4-inch legs and a thickness of 1/4 inch.

I-BEAM

An I-beam is used to span openings and for other structural purposes. I-beams are sized by their height. Different weights are

TABLE 4.8
Lintel Sizes

Wall Thickness, in.	Span			
	3 Ft		4 Ft* Steel Angles	5 Ft* Steel Angles
	Steel Angles	Wood		
8	2-3×3×¼	2×8	2-3×3×¼	2-3×3×¼
		2-2×4		
12	3-3×3×¼	2×12	3-3×3×¼	3-3½ × 3½ × ¼
		2-2×6		

Wall Thickness, in.	Span		
	6 Ft* Steel Angles	7 Ft* Steel Angles	8 Ft* Steel Angles
8	2-3½ × 3½ × ¼	2-3½ × 3½ × ¼	2-3½ × 3½ × ¼
12	3-3½ × 3½ × ¼	3-4×4×¼	3-4×4×¼

* WOOD LINTELS SHOULD NOT BE USED FOR SPANS OVER 3 FEET SINCE THEY
BURN OUT IN CASE OF FIRE AND ALLOW THE BRICK TO FALL.
COURTESY OF THE BRICK INSTITUTE OF AMERICA

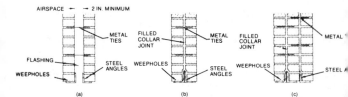

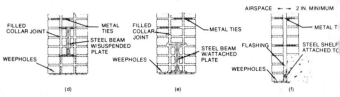

Fig. 4.9 Types of structural steel lintels   Courtesy of Brick Institute of Ame

available for each size. Industry nomenclature describes the size and shape of the beam and its weight per linear foot. A standard 12-inch I-beam weighing 35 pounds per linear foot is designated "S 12 × 35."

A wide-flange style is available and more commonly used. Wide-flange beams have greater bearing surface and are available in more weights than standard sections. The designation is similar to that used for the I-beam, except that "W" indicates that it is a wide-flange. A 14-inch wide-flange beam weighing 87 pounds per foot is designated "WF 14 × 87."

REINFORCING BAR

Reinforcing bar, also called re-bar or re-rod, is essential to increasing the tensile strength of concrete and masonry structures. Re-bar is commonly used in these applications: 1) slabs, 2) bond beams, and 3) grouted collar joints.

Reinforcing bar is "deformed" with a pattern to help concrete and grout gain a better grip on the bar. Special shapes may be prepared by steel suppliers or bent on the job. Reinforcing bar should be bent cold, never hot.

Reinforcing rod is sold in various diameters. Each bar may be stamped on the side with a number to indicate diameter in eighths of an inch. A bar marked "6" is 6/8 inch, or 3/4 inch, in diameter. The rod must have adequate diameter to do the job but still be small enough to be surrounded by the grout, concrete, or mortar. Size and placement of reinforcing bar is specified in blueprints.

BLOCK REINFORCEMENT

Reinforcing rod is used to create strong columns and beams in block structures. A wall brac, or spider, is a small wire jig sometimes used to ensure that vertical rods remain located in the cavity while the grout is pumped in. A cradle positioner is used for the same purpose in bond beams (see *Joint Reinforcing*, p. 84).

CONCRETE REINFORCEMENT

Reinforcing rods and mesh are used to increase the tensile strength of concrete structures. The reinforcement must be fully enclosed by the concrete; several types of bolsters can be used to lift the reinforcing rod off the bottom if needed.

Special shapes and configurations are available for large jobs as specified by the architect or engineer. The ends of rods are frequently bent into a hook shape to increase the bond with the concrete and prevent it from creeping along the reinforcement. Bars and mesh must be free of oil or mud; otherwise, the concrete will not bond to them.

TABLE 4.9
Sizes of reinforcing rod

Bar Designation	Fractional size (inches)	Unit weight (pounds/foot)	Diameter (inches)
3	3/8	0.376	0.375
4	1/2	0.668	0.500
5	5/8	1.043	0.625
6	3/4	1.502	0.750
7	7/8	2.044	0.875
8	1	2.670	1.000
9	1-1/8	3.400	1.128
10	1-1/4	4.303	1.270
11	1-3/8	5.313	1.410
14	1-3/4	7.65	1.693
18	2-1/4	13.60	2.257

STEEL PLATE

Steel plate is used for reinforcing and extending other steel products. Plate may be welded to the top of a post or the bottom flange of an I-beam to provide a bearing for masonry units. Steel plate is designated by thickness and width.

STRAP ANCHOR

Strap anchors are used to tie intersecting walls together, particularly when some movement is required. Strap anchors for block intersections are commonly made 1/4 inch thick, 1-1/4 inch wide, and 24 inches, 28 inches, or 30 inches long (see *Walls, intersections*, p. 107).

JOINT REINFORCING

Joint reinforcing is laid into mortar joints to increase lateral strength and reduce chances of failure due to earthquake, wind, and other loads. Reinforcing material with high tensile strength will not eliminate cracking due to shrinkage, but it will distribute the stresses, resulting in a number of small cracks instead of a few large ones.

Three types of joint reinforcement are used for walls: truss type, ladder type, and tab type. Truss type offers greater strength than ladder type, but is more expensive. For composite walls of block and brick, ladder-type reinforcing is preferred because it allows more differential movement. Reinforcing for concrete block should provide one side wire for each face shell, so composite walls require either three-wire reinforcing or tab-type reinforcing. Joint reinforcement may be deformed to allow the mortar a greater grip; several finishes are available. Corrosion-resistance is desirable in most applications.

Preformed inside and outside corners are available, but they are seldom used because reinforcing is easy to cut and bend in the field. Reinforcement is commonly used 16 inches O.C. vertically in walls. Openings need special attention; reinforcement is generally used 8 inches above and below a window, and 8 inches above a door opening. Reinforcing is specified on blueprints (see Fig. 4.10).

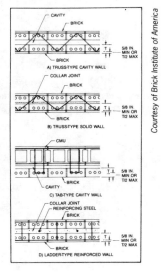

Courtesy of Brick Institute of America

Fig. 4.10   Joint reinforcing

STONE MASONRY

Stone, the original mason's material, still produces a strong, attractive wall. Stone is challenging to work with and generally slower to lay than manufactured masonry units.

Each of these three basic types of rock can be used as building stone:

1. *Sedimentary rock* is formed by the gradual accumulation of particles. This category includes limestone and sandstone.

2. *Igneous rock* is formed at high temperature deep within the earth. The major igneous building stone is granite, which can be sawed into thin slabs as a veneer or laid in rough form for walls.

3. *Metamorphic rock* results when extreme heat and pressure deform other types of rock. For builders, this category includes marble (from limestone), quartzite (from sandstone), and slate (from shale).

From a mason's point of view, stone should be evaluated for these characteristics:

- *Strength. Most stone has adequate compressive strength but some types tend to split.*

- *Hardness. Most stone is sufficiently hard. Do not use soft stone in areas with heavy traffic.*

- *Durability. Depending on the application, stone may need to resist moisture, dust, frost, and fire. A durable stone will need less maintenance than a nondurable one.*

- *Workability. The stone should be available in a variety of sizes to reduce the need for cutting. If not, it should be easy to cut.*

- *Color and grain. Mostly aesthetic concerns.*

- *Porosity. May affect how well the stone will withstand frost action, marking, and staining.*

- *Texture. Affects workability and cost of construction. Fine textured stone is easier to work because it cuts more readily than coarse-textured stone.*

Although economics often dictates the use of local stone, high-quality stone may be shipped a long way; check masonry supply firms.

Either ashlar or rubble masonry can be laid up "coursed"—with courses like a brick wall, or "random"—without courses. Flagstones, such as slate, are used to pave floors.

Sawed stone is used for details, such as copings, lintels, sills, quoins, jambs, arches, and steps, often in structures built of other masonry materials.

ASHLAR

Ashlar masonry has courses and the faces of the stones are squared. The three categories of ashlar walls are random ashlar, irregular coursed ashlar, and regular coursed ashlar.

ARTIFICIAL

Artificial stone is formed of portland cement, aggregates, and pigments in an attempt to resemble natural stone. The finished stone wall can be as thin as 1-3/4 inches. Artificial stone is often very light and sometimes may be veneered over a mortar coat on the base wall, with no foundation needed. Due to its regularity and ease of cutting, artificial stone lays up much faster than rubble.

For exterior walls, a 4-inch clearance is suggested above grade to prevent staining. Special pieces are used for corners. Capping is needed to prevent water from seeping behind or running down the surface of exterior walls.

Artificial brick, a variation of artificial stone, is made in sections about 1/2 inch thick. The brick is bonded to a plaster base coat, with no need for a foundation or masonry backing wall.

RUBBLE

Unshaped stone, either rubble or fieldstone, presents a challenge to the mason, who must take great care to achieve an attractive wall and a strong bond. Rubble stone should be chosen for color and shape. Rubble can be random—laid with no attempt to make courses. If the stone is relatively regular, it can be laid in rough courses, called "coursed rubble." Sometimes, rubble is split to expose a fresh face before laying.

In rubble work, as in almost all masonry, the bond must be "broken," so head joints do not line up in successive courses.

TERRA COTTA

Terra cotta, literally "cooked earth," is a clay building material that is frequently glazed. Most often used now in restoration work, terra cotta can be formed into complex shapes not feasible with other masonry materials. Terra cotta treatments are commonly found around entryways and in decorative trim courses. When it is needed for remodeling and restoration, molds for replacement parts can be taken from existing features.

TILE

Fired clay may be formed into many types of tile: both facing and structural. Structural tile is now rare, having been largely replaced by concrete block. Tile parts are named like block parts: the outer sides are shells, the members joining the shells are webs, and the hollow spaces are cells. Tile may be laid either on end, called end construction, or on the side, called side construction. Mortar joints are generally about 1/2 inch thick.

LOADBEARING WALL

Loadbearing tile is used for loadbearing or nonbearing walls and is best above grade. Codes usually restrict this tile to four stories, or 40 feet in height. This tile may stand alone or serve as a backing for veneers of other material.

PARTITION

A non-bearing tile in thicknesses of 2, 3, 4, 6, 8, 10 and 12 inches. It is generally 12 inches square in nominal dimension.

BACKUP

Similar to loadbearing tile, but intended only for use as a backup wythe.

FURRING

These tiles provide a space for insulation, a surface for plastering on the inside of the wall, and an increase in a wall's water resistance.

FIREPROOFING

These tiles are used to protect structural steel from the heat of fire as required by building codes. Fireproofing tiles are shaped to fit structural steel. This can also be done with furring, partition, or wall tiles.

FLOOR

Floor tile is used and made in several sizes and shapes:

1. Floor tiles are used for laying over concrete floor. Many types are used, depending on the application.

2. Ceramic mosaic tiles are small pieces arranged in patterns, with a nominal thickness of 1/4 inch. The tiles are bonded to backings in large pieces to simplify installation.

3. Quarry tile is dense, unglazed tile made in squares, oblongs, and other shapes in a thickness ranging from 1/2 inch to 3/4 inch.

4. Paver tiles are intended for slightly lighter use, and range from 3/8 inch to 5/8 inch.

5. Packing house tiles are used in industry, ranging from 1/4 inch to 1-5/8 inch thick.

6. Slip-resistant finishes are available on some types. Trim tiles are available for floor moldings and stair nosings in some cases.

STRUCTURAL CLAY FACING

These tiles have either a rough or finished face and are suitable for exterior and interior applications. Standard tiles are designed for general applications, while special-duty tiles have greater resistance to moisture and impact due to their thicker shells and webs. The tiles can also be glazed on one or two sides, called structural glazed facing tile.

CERAMIC

Ceramic, decorative tile is a specialty material that is used strictly for esthetic purposes and has no structural role. Tilesetters perform most tile laying, but masons may be called upon to lay hearths, walls, and counter tops of ceramic tile. This tile is generally set into a mortar or adhesive and the voids filled afterward with grout of the desired color.

WALL TIES

Wall ties are used to join a veneer wythe to a wood- or metal-studded wall or to replace masonry bonders in a two-wythe wall. The nature of the tie will depend on the nature of the backing and the facing wythe.

Wall ties must meet these standards:

1. Easy to install.
2. Securely embedded in the bed joints.
3. Stiff enough to transfer lateral loads.
4. A minimum amount of mechanical play.
5. Resist corrosion.

Some adjustable ties are designed to allow differential movement between the wythes. Strap anchors and other devices serve as wall ties and are considered in this section. Other devices also can accomplish some of the tasks of wall ties (see also *Anchors*, p. 154; *Reinforcement, Joint*, p. 84).

Specific requirements vary, but it is common to use one anchor for each 4-1/2 square feet of wall. The usual practice is to avoid placing ties directly above each other, to spread out the reinforcing action, but this may be impossible when the ties are affixed to the studs. Wall ties, and especially the masonry bonders they replace, may provide a path for water to travel from the exterior to interior wythes.

Ties and their placement are specified in blueprints. Two common types are Z- and rectangular-shaped, corrosion-resistant ties of 3/16-inch rod. Rectangular ties are suggested for either hollow or solid units, while Z-ties are for units with no more than 25 percent cores. A third type, the corrugated tie, is only suitable for low residential veneers because 1) it has little strength for tying the wythes together, 2) its shape promotes water travel, and 3) it is susceptible to corrosion.

Tab-type and other forms of joint reinforcing eliminate the need for wall ties because the facing wythe is bonded to the same piece of reinforcement as the backing wythe.

Wall ties may have an indent in the center called a "drip." The drip must face down so any condensation drips into the wall cavity, not down the wythe (see Fig. 4.11). However, the drip reduces the tie's lateral strength, which may indicate a greater number of ties.

ADJUSTABLE TIES

Adjustable wall ties are indicated when:

1. The tops of units in the two wythes do not coincide.
2. Large vertical differential movement is expected.
3. Walls have nonmasonry backup.

Adjustable ties are flexible and less subject to damage during construction than other types of tie. In addition, they allow a building to be rapidly enclosed. However, they should be used with care, because they may suffer drastic reductions in strength when set at maximum adjustment. Adjustable ties must be placed relatively accurately to avoid being used at maximum adjustment.

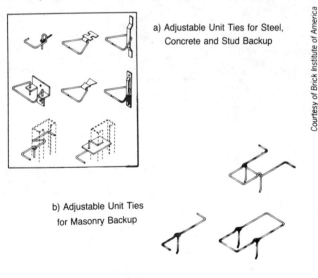

a) Adjustable Unit Ties for Steel, Concrete and Stud Backup

b) Adjustable Unit Ties for Masonry Backup

Courtesy of Brick Institute of America

Fig. 4.11   Wall ties and anchors

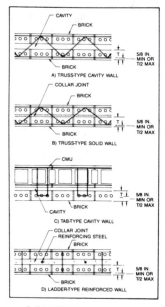

A) TRUSS-TYPE CAVITY WALL

B) TRUSS-TYPE SOLID WALL

C) TAB-TYPE CAVITY WALL

D) LADDER-TYPE REINFORCED WALL

c) Joint Reinforcement Details

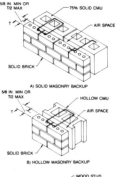

A) SOLID MASONRY BACKUP

B) HOLLOW MASONRY BACKUP

C) WOOD STUD BACKUP

d) Unit Tie Details

Fig. 4.11   Wall ties and anchors (cont.)

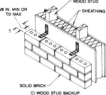

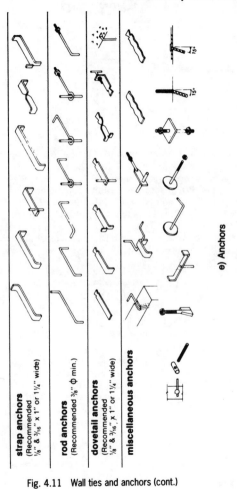

strap anchors
(Recommended
1/8 & 3/16" x 1" or 1 1/4" wide)

rod anchors
(Recommended 3/8" φ min.)

dovetail anchors
(Recommended
1/8 & 3/16", x 1" or 1 1/4" wide)

miscellaneous anchors

e) Anchors

Courtesy of the Indiana Limestone Institute

Fig. 4.11   Wall ties and anchors (cont.)

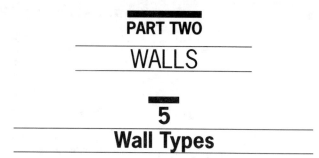

PART TWO

WALLS

5
Wall Types

The types of bearing masonry walls include: foundation and basement, composite, and cavity. The types of nonbearing walls include serpentine, retaining, garden, and veneer. Each type of wall is suited to certain applications and has particular requirements for correct, efficient construction.

Fig. 5.1 can be used for identifying many types of brick walls, and Table 5.1 contains information about the properties of these walls. Characteristics of composite walls made of block and brick will vary somewhat from those shown.

CAVITY

Cavity walls, also called hollow walls, are multiple-wythe walls with a gap of at least two inches between two of the wythes. Cavity walls are specified for their superior resistance to moisture and heat and sound transmission. The separation of the wythes also allows a great deal of differential movement without cracking. It can also prevent efflorescence caused by salts from the backing wythe migrating into the facing wythe. The cavity may be insulated with pourable or rigid material.

Cavity walls can be bearing or nonbearing. The usual practice is to bear the weight of other structural elements on the inner wythe only. For heavy loads, the interior wythe can be made wider than the facing wythe.

TABLE 5.1
Wall Properties

Wall Number		1	2	3	4	5	6	7	8	9	10	11	12
Material Quantity[a] (per 100 sq ft)	Brick	675	450	960	1350	960	960	300	1350	1350	450-4" 450-6"	2025	600
	Mortar (cu ft)[b]	5.5	7.9	11.3	14.1	8.1	8.1	7.8	10.9	10.9	12.9	22.7	7.4
	Grout (cu yd)	—	—	—	—	—	0.62	—	—	0.74	—	—	1.36
	Metal Ties[c]	—	—	23	23	23	34	—	23	34	23	—	34
	Reinforcement (lb)[d]	—	—	—	—	—	65	—	80	80	—	—	95
U Value[e] (BTU per sq ft-hr-deg Fahr)	Uninsulated	0.76	0.68	0.60	0.54	0.38	0.54	0.54	0.36	0.49	0.33	0.42	0.44
Wall Weight (lb per sq ft)	Unplastered	40	60	65	80	60	85	80	75	100	95	120	125
STC (db)	Unplastered	45	51	51[h]	52	50[h]	52[h]	52	50	59	59[h]	59	59[h]
Fire Resistance[f] (hr)	Unplastered	1	2	3	4	3[g]	4	4	4	4	4	4	4

COURTESY OF THE BRICK INSTITUTE OF AMERICA

[a] WASTE IS NOT INCLUDED, AS THIS WILL VARY WITH THE JOB. A WASTE FACTOR OF 5 PERCENT IS FREQUENTLY APPLIED FOR MASONRY UNITS AND 10 PER CENT TO 20 PER CENT FOR MORTAR AND GROUT.

[b] ASSUMED MORTAR JOINT THICKNESS 3/8 IN.

[c] BASED ON ONE METAL TIE FOR EACH 4-1/2 SQ FT OF WALL AREA FOR NON-REINFORCED WALLS, AND ONE METAL TIE FOR EACH 3 SQ FT FOR REINFORCED WALLS (HIGH-LIFT GROUTING ASSUMED).

[d] BASED ON MINIMUM REINFORCEMENT OF 0.002 TIMES THE CROSS-SECTIONAL AREA OF THE WALL.

[e] CORRECTED FOR A 15-MPH WIND OUTSIDE AND STILL AIR INSIDE.

[f] NON-COMBUSTIBLE OR NO MEMBERS FRAMED INTO WALL.

[g] WALL TESTED HAD 3-IN. (ACTUAL) THICK WYTHES AND WAS 9 FT IN HEIGHT. MORTAR FILL (1 FT IN HEIGHT) WAS ALSO PLACED BETWEEN WYTHES AT TOP OF WALL.

[h] ESTIMATED QUANTITY.

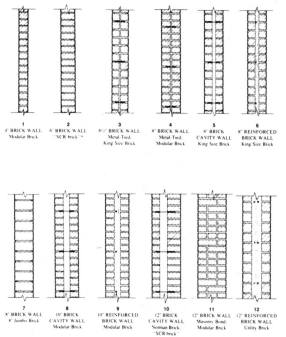

Fig. 5.1   Wall type identification

If the wall is supported on shelf angles, provide horizontal pressure-relieving joints below each angle to allow movement. Allow a small space for expansion at each end of the shelf angles. Expansion joints should be installed on one or both sides of an outside corner if walls are longer than 50 feet, and in other positions as dictated by the architect (see Fig. 5.2).

Flashing. Rigid flashing material is bent into the proper shape and inserted into both wythes of the cavity wall (vinyl can be placed without bending). If wind-driven rain penetrates the exterior

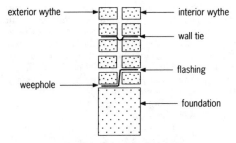

exterior wythe → ← interior wythe

wall tie

flashing

weephole →

foundation

Fig. 5.2 Parts of a cavity wall

wythe, it will drip to the flashing and be diverted through weepholes to the exterior. Flashing must be set accurately, and you should avoid disturbing flashing when working around it. Flashing can break the mortar bond and is sometimes specified to do this, such as in the instance of two adjacent building elements with different rates of thermal expansion.

Weepholes. Weepholes allow water that accumulates on the flashing to exit the wall. Weeps can be formed by inserting sash cord or a similar material in specified head joints, or by leaving the head joints dry. Weeps are usually placed 2 feet or 3 feet O.C. Weeps must not be obstructed by mortar droppings in the cavity while you build the wall. You can prevent clogging by placing rubble in the cavity behind each weep. If granular insulation is used, the weeps should be screened to prevent clogging.

Keep the cavity clean. The cavity must be kept clear for the wall to perform properly. This procedure will also simplify the placement of insulation. Bevel mortar as you apply it, so it is thicker on the side away from the cavity. It is not always necessary to make the joints full on the inside. The general practice is that joints are acceptable if they are full to within 1/4 inch of the inside surface. Hold the trowel flat as you cut off mortar from joints, so you can catch the mortar and prevent it from dropping into the cavity.

Use a length of 1-inch × 2-inch lumber suspended from wire inside the cavity as a mortar tray to catch droppings. Periodically remove the tray and dump the mortar from it.

Wall ties. Wall ties are generally specified at a rate of one per 4.5 square feet of wall surface. Place ties at 16 inches O.C. vertically and 36 inches O.C. horizontally. Where possible, stagger the ties so they are not directly above and below each other. The maximum vertical distance between ties should be 24 inches, and the maximum horizontal distance 36 inches. Place additional ties within 12 inches of openings, at no more than 36 inches apart. Place extra ties directly below all floor joists. If the cavity is between 3-1/2 inches and 4-1/2 inches wide, use one tie per 3 square feet.

Bond wall ties by placing mortar on the wall and then pressing the tie into the mortar. Wall reinforcement that bonds multiple wythes can be used to replace individual wall ties. Make sure to lap the reinforcement at least 8 inches and provide continuous reinforcement around corners. Reinforcement should be discontinuous at movement joints.

Some ties have a "drip," a U-shaped projection in the middle that forces water to drip into the cavity, instead of down the wythes. Face the drip downward.

Insulation. Make sure to fill cavities below window sills before you cover them with subsequent work. The chosen insulation should not prevent water from draining inside the cavity. Rigid insulation is installed in the cavity as the wall is constructed. Bond the boards to the inside of the backing wall with mastic tape or anchors. You may have to cut the insulation in 24-inch strips for easy placement. Granular insulation, such as perlite, vermiculite, or expanded foam, can be poured at each change in the scaffold or from the top of the wall. Granular insulation should be designed to prevent settling over time, and be pourable in lifts of at least 4 feet. Do not rod or tamp granular insulation.

CAVITY WALL PROCEDURE

This procedure is for a wall laid starting with the outer wythe. If you prefer, it can be adapted to lay the inside wythe first, place the flashing, and then lay the outer. This method reduces the chance of disturbing the flashing while laying the wall.

1. Lay out the wall. Clean the foundation top. Mark the inside and outside corners with a pencil on the foundation. Snap lines for the outside of the exterior wythe and the inside of the

interior wythe. Check that the lines are parallel for their entire length.

2. Lay out the first exterior course dry. Check head joint spacing and mark them when they are correct. Calculate cutting requirements and cut enough units for the wall, or plan to cut them as you go.

3. Prepare the flashing. Check specifications for how far it must penetrate each wythe. Measure and cut to width and length. Bend the flashing into the proper form. (If you are using vinyl, no bending is needed.)

4. Lay up the first exterior course. If a bond break will be used between the foundation and the course, install it now. Lay a shallow bed of mortar, place the flashing into the bed, and place more mortar on top. Smooth this mortar and place the units into position.

5. Form weepholes in the head joints at specified intervals. Place sash cord temporarily in the head joint or leave the head joint dry.

6. Lay the leads for the exterior course and continue laying the wall.

7. Lay the first interior course dry if necessary to establish the bond. Then mortar the first course in place underneath the flashing. Take care not to disturb the flashing.

8. Lay some mortar on top of the first interior course and press the flashing into it. Then spread more mortar and place the units for the second interior course.

9. Insert a mortar tray into the cavity and hang its wires over the side. Continue laying the courses as specified. Remove and empty the mortar tray periodically. Install rigid insulation before placing wall ties. Lay wall ties and joint reinforcement as specified. Use a beveled mortar placement to reduce mortar droppings.

10. Tool the joints per specification. Brush scrap mortar from the wall. Check that the cavity remains at proper width and both wythes are true, plumb and level.

11. Remove sash cords from the drips when the mortar sets. Clean the wall if needed.

Above-grade wall equipment and supplies. Use this checklist to prepare estimates and order materials:

Information needed:
Blueprint and specification
Brick or stone pattern
Weephole locations
Schedule for construction
Joint spacing and type
Cartage distances

Preparation:
Water and power
Site grading
Foundation or footings
Layout reference points

Materials:
Brick, block and stone
Insulation
Anchors and angle bolts
Joint reinforcement
Masonry nails
Angle iron, plates, and
  I-beam
Chase equipment
Cleaning supplies
Fuel
Admixtures
Mortar and color for each
  material
Wall ties
Lintels and sills
Flashings
Control joints
Caulking
Sealants

For stucco:
Control joints
Mesh
Corner beads
Applicator (if needed)

Equipment:
Scaffold
Vibrator and pump for
  grout
Covering for walls and
  materials
Cold–weather enclosure
  for workers
Mixer
Saw or splitter
Heating equipment for
  cold weather
Safety equipment
Sealant applicator
Cleaning equipment

COMPOSITE

Composite walls have multiple wythes of different materials, generally bonded together with masonry or steel. Composite walls are loadbearing, exterior walls with cold weather- and fire-resistance, and low maintenance requirements. However, they have less insulating value than cavity walls. A common type of composite wall combines the economical construction of concrete block on the interior and the desirable appearance of brick on the exterior.

The key to a composite wall is to ensure a good structural bond between the two wythes. This bond is accomplished with headers (masonry bonders), reinforcement, wall ties, or parging of the collar joint. Masonry headers are prohibited by some building codes because they transmit too much heat.

1. When using headers, carefully figure at what course interval they will be needed. Headers are frequently used every five to seven courses. A header block that can be used with brick headers in 12-inch walls is shown in Fig. 5.3 The recess in the block accepts the header and preserves the coursing and bond. Take care when placing header blocks on top of freshly laid headers, as the blocks can lift the headers from the exterior wythe and ruin the structural bond.

2. Wall ties are often used every six courses. Place the drip facing down, so water will drip between the wall, not down the inside of the wythes.

3. Special reinforcement can reinforce the joints and tie the two wythes together with a single piece.

4. On some low composite walls, parging in the collar joint is the only bond needed between the wythes.

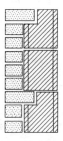

Block can be reversed for other bonding patterns

Fig. 5.3 Composite wall with masonry headers

Either the facing (exterior wythe) or the backing (interior wythe) can be laid first. Parging is frequently specified between the wythes to increase resistance to water. Aside from the method used to tie the wythes together, construction of composite walls is similar to construction of cavity walls. In cold climates, composite walls are less common than cavity walls, which offer greater insulation.

FOUNDATION AND BASEMENT

Foundation walls are those that support the first floor above grade level. Foundation walls that enclose usable space are called basement walls. The materials used in foundation and basement walls must be able to resist weathering and moisture. All exterior joints must be properly tooled with a concave or V-joint to resist moisture. The exterior may be parged, or asphalt or a membrane may be applied to prevent water entry. Tool interior joints per specifications.

Foundation walls vary as much as the buildings they support. Instructions are given here for a typical house foundation of concrete blocks:

1. Pour footings on stable soil below the frost line (see *Footings,* p. 233).

2. Inspect the footings to be sure they are level and in the proper location. Find the actual foundation corners and pound nails in the green concrete to mark them. Then snap chalk lines for the wall lines.

3. Lay the units dry to check the bond (or use a rule to make sure the layout conforms to desired spacing). Lay the first course in a full bed of mortar.

4. Build up leads at the corners, using face-shell bedding on all courses above the footing (see *Leads, Building,* p. 217).

5. Lay stretchers between the leads, using face-shell bedding. Maintain the proper course height.

6. Provide cavities for angle bolts. If 15-inch bolts are being used, place wire mesh under the second course from the top. For 8-inch anchors, place mesh under the top course. Cut mesh 1 inch narrower than the block's actual width.

7. Finish laying the top courses. Insert angle bolts in the required cores and brace them in a vertical position at the correct height. Grout all cores holding angle bolts. Cut off all grout flush with

the top of the blocks. (In some applications, especially when the foundation is thicker than the above-grade wall, a course of solid blocks is used to finish the wall.)

8. Brace the walls if necessary until other building elements are in place to support them.

9. If you are assigned the task of installing the sills, drill holes in the proper places in the sills and attach them to the wall after the grout has cured. In wooden sills, it is good practice to drill recesses so the washers and bolts do not protrude above the sill. Then bolt the sill down, making sure to locate it exactly.

REINFORCED FOUNDATION WALLS

Horizontal reinforcement and low-lift grouting techniques can be used to reinforce a foundation wall (see *Concrete block construction, bond beam*, p. 209; and *Grouting*, p. 240). A bond beam is commonly used to strengthen the top course. Use lintel blocks for this course (cut holes in the bottoms to allow installation of 15-inch angle bolts). Lay the lintel block and place jigs to hold the horizontal reinforcement in location in the bond beam. Then brace angle bolts with blocks of 2 × 4 lumber. Puddle the mortar or grout into the lintel block course and allow it to set. Then remove the blocks.

Several techniques can be used to integrate a block foundation wall with concrete floors. Solid blocks may be used to maintain the 8-inch bonding. Mesh may be placed atop the top course and the slab cast in place. Solid top units may be laid as the last course before a cast-in-place concrete floor.

Foundation equipment and supplies. Use this as a checklist when preparing estimates and ordering materials:

Information needed:
Blueprint and specification   Cartage distances
Schedule for construction

Other contractor's responsibilities:
Excavation   Footings
Backfilling

Preparation:
Water and power   Excavation
Layout points   Footings
(benchmarks)

Materials:

Masonry cement and sand
Admixtures
Re-rod
Insulation—poured and rigid
Joint reinforcement
Masonry nails
Door and window frames
Flashing
Caulking and joint filler
Fuel for heater
Anchor bolts

Concrete or grout for reinforcing
Block—standard and special (corner, solid, and pilaster)
Wall ties and fasteners
Lintels and sills
Angle iron and I-beam
Control joints
Sealant and/or parging supplies

Equipment:

Scaffolding and planks
Grout pump
Sealant applicator
Heater for cold weather
Covering for walls and materials

Bracing
Mortar mixer
Saw
Cold-weather enclosure for workers
Safety equipment

GARDEN

Garden walls are used for security, privacy, and landscaping purposes. Three major types are built, with many variations: straight, pier and panel, and serpentine.

Garden walls generally have two wythes and rest on footings set below the frostline. Subgrade foundations can be of concrete block, but in areas with significant soil moisture and freezing conditions, poured concrete is best.

Use Type S mortar for garden wall construction. Mortar can be souped up with water to make grout for piers and other reinforced areas. Copings can be a rowlock course of brick, natural stone, terra cotta, slate, or metal. If the coping is made of a different material than the wall, check that its thermal expansion rate is similar to that of the wall material. Copings should be weathertight, extend at least

1/2 inch past the wall and have a positive drip to prevent rainwater from running back down the wall. Anchor copings to the wall with anchor or angle bolts. Some masons place flashing under the coping to prevent water from penetrating the wall, especially in areas with severe weather.

STRAIGHT WALL

Straight walls are built with no piers, and they have little resistance to turnover. Long sections of straight wall are subject to damage by wind pressure and possible vehicle impacts. Wind pressure must be considered when calculating wall thickness and design. 20 pounds per square foot (psf) is the specified design wind load for garden walls in many building codes. (Gusts of 80 mph produce 20 psf pressure.) Design wind loads vary by location—consult local authorities. A rule of thumb is to limit the height to 10 wall thicknesses if the wind pressure has a maximum of 20 psf. Thus, a wall constructed of 8-inch blocks must not exceed 80 inches high.

PIER AND PANEL WALL

Pier and panel walls combine the economical coverage of 4-inch panels with the support of thicker piers. The piers should be at least as tall as the panels.

Piers must rest on footings below grade. Three types of footings are possible: 1) Uniform footings have the same width for their entire length. 2) Proportional footings are wider under the piers than under the panels. 3) Separated footings support only the piers. Panels laid on separated footings must be braced during layup. It is best to reinforce these panels because they have no direct support under their centers after the bracing is removed.

The spacing of reinforcing in pier and panel walls depends on wind load and panel length. Two reinforcing locations are required as indicated in Table 5.2: 1) Panel wall steel, consisting of pairs of reinforcing bars placed in joints at the specified spacing; 2) Pier reinforcing steel, consisting of pairs of bars placed vertically in each pier, spaced about 2 inches apart.

The pier footings must extend below the frost line or to the depth indicated in Table 5.2, whichever is deeper (see *Piers,* p. 266, for procedures for building piers).

TABLE 5.2
PIER AND PANEL WALL SPECIFICATIONS
Panel Reinforcement

Wall Span, ft	Maximum Spacing of Two No. 2 Bars		
	Wind Load, psf		
	10	15	20
8	3'— 8"	2'— 4"	1'— 10"
10	2'— 4"	1'— 7"	1'— 2"
12	1'— 8"	1'— 1"	10"
14	1'— 3"	10"	7"
16	11"	7"	6"

TABLE 5.2 (cont.)
Pilaster Reinforcing Steel

Wall Span, ft	Wind Load, 10 psf Hall Height, ft			Wind Load, 15 psf Wall Height, ft			Wind Load, 20 psf Wall Height, ft		
	4	6	8	4	6	8	4	6	8
8	2#3	2#4	2#5	2#3	2#5	2#6	2#4	2#5	2#5[1]
10	2#3	2#4	2#5	2#4	2#5	2#7	2#4	2#6	2#6[1]
12	2#3	2#5	2#6	2#4	2#6	2#6[1]	2#4	2#6	2#7[1]
14	2#3	2#5	2#6	2#4	2#6	2#6[1]	2#5	2#5[1]	2#7[1]
16	2#4	2#5	2#7	2#4	2#6	2#7[1]	2#5	2#6[1]	2#7[1]

[1] REQUIRES 16-IN. PILASTERS.

TABLE 5.2 (cont.)
Required Embedment of 15-In. Pier Foundation

Wall Span, ft	Wind Load, 10 psf Wall Height, ft			Wind Load, 15 psf Wall Height, ft			Wind Load, 20 psf Wall Height, ft		
	4	6	8	4	6	8	4	6	8
8	2'-0"	2'-6"	2'-9"	2'-3"	2'-9"	3'-3"	2'-3"	3'-0"	3'-9"
10	2'-0"	2'-6"	3'-0"	2'-3"	3'-0"	3'-9"	2'-6"	3'-3"	4'-3"
12	2'-3"	2'-9"	3'-3"	2'-6"	3'-3"	4'-0"	2'-9"	3'-9"	4'-6"
14	2'-3"	3'-0"	3'-6"	2'-9"	3'-6"	4'-3"	3'-0"	4'-0"	4'-9"
16	2'-3"	3'-0"	3'-9"	2'-9"	3'-9"	4'-6"	3'-3"	4'-3"	5'-3"

INTERSECTIONS

Intersections are locations where walls meet other walls. T-intersections are perpendicular junctions where one wall ends against another. Continued intersections are junctions where two walls cross and do not terminate. Intersections can be constructed with a tight bond, allowing no movement, or a loose bond, which allows some movement. The type of intersection will depend on the application and blueprints. In general, if both walls at an intersection are loadbearing, use strap anchors or masonry bonding for a rigid connection. If one or both walls are nonbearing, then wall ties, wire mesh or wire lath provides sufficient bonding.

MASONRY BOND

In a true masonry-bonded intersection, units from one wall are embedded in the other wall. Each unit making the bond should bear at least 3 inches on the unit below. The details of the masonry bond depend on the size of units being used. Use a dry layout to plan the bond.

Sometimes units must be cut to make the proper bond, especially when using blocks other than 8-inch nominal thickness. For strength, try to cut at least 2 inches from any block in a masonry-bonded intersection. Note that 4-inch pieces are used in the continued 4-inch block intersection shown in Fig. 5.4. On the first course, one wall is uninterrupted and the other butted against it, with cut units used to preserve the masonry bond. On the second course, this pattern is reversed.

A masonry bond for an 8-inch T-intersection uses butt joints for even-numbered courses and a 4-inch bond for odd-numbered courses. During dry layout, try to lay the units so the intersecting wall lines up with the end of a full block in the perimeter wall. Strap anchors, shown in Fig. 5.5, or wire lath can be used to strengthen a masonry-bonded intersection, especially when the intersecting wall is laid after the perimeter wall.

A masonry-bonded 8-inch T-intersection uses sawed blocks in both walls in odd-numbered courses. The block in the intersecting wall is cut to 12 inches long.

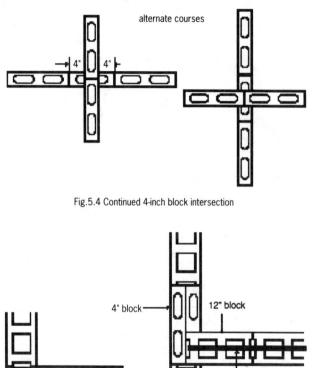

alternate courses

Fig.5.4 Continued 4-inch block intersection

4" block

12" block

use strap anchor when laying walls separately

Fig. 5.5 8-inch T-intersection—masonry bond

FLEXIBLE INTERSECTIONS

Walls that must shift in one direction and be rigid in the other require a connection with the proper combination of support and movement. This can be done by grouting steel strap anchors into cores at a spacing no greater than 4 feet O.C. vertically. Walls connected with strap anchors can be laid simultaneously or separately, using slightly different procedures. Use the following procedure to install strap anchors in walls laid simultaneously (see Fig. 5.6):

1. Lay up both walls to one course below anchor height. Install wire mesh under the core in which you will insert the downpointing hook of the anchor.

2. Lay the next course. Grout the core with wire mesh below it. Place the anchor so one hook is buried into the grout. Lay wire mesh under the other end of the anchor to support grout that will be poured after the next course is laid.

3. Lay the next course and fill the remaining core with grout.

4. Rake out mortar in the joint between the intersecting walls to the proper depth. Do this for each course as the work progresses.

5. Fill the raked joint with caulking if required for weather resistance or appearance.

If the perimeter wall must be built first, embed the anchors in the required positions as in the previous procedure. When building the cross wall, tilt and slide units under the anchors without disturbing them. Then pour grout into the core.

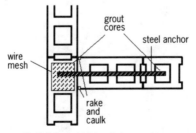

Fig 5.6 Control joint and strap anchor

BOND BEAM

Bond beams can be used to reinforce intersections in block walls, especially in areas subject to earthquake, blast or high winds. Use the following procedure to lay bond-beam intersections (see *Concrete block construction, bond beam,* p. 209):

1. Lay both walls to the course below the bond beam. If hollow bond beam units are used, install lath on top of the next-to-last course to hold the grout.

2. Lay bond beam units on both walls. Cut one unit in the perimeter wall where reinforcing bars from the intersecting wall will cross into the perimeter wall.

3. Install re-bars as specified. Note that two full bars are used in the perimeter wall in addition to the bent bars joining the walls together.

4. Grout the bond beam and vibrate or puddle the grout into place.

BRICK

Intersecting, bearing 8-inch brick walls can be joined with a full masonry bond as shown in Fig. 5.7. The size of any bats used depends on the size and proportions of the bricks. Plan the bond to minimize the number of cuts required. Both walls are generally constructed at the same time. If the perimeter wall is built first, leave voids so the intersecting wall can be keyed into place later on.

Butted intersections may be used for nonbearing walls. Join the walls with Z-shaped wall ties at specified vertical intervals. To allow some movement, rake head joints at the intersections. Caulk this joint to improve weather resistance or appearance.

Brick walls joined in continuing intersections are generally joined by full masonry bonds. If possible, lay out courses so cut units are not needed.

PARAPET WALL

The parapet wall is subject to cracking, displacement, and leaking. The Brick Institute of America suggests that the design shown in Fig. 5.9 for a parapet on a cavity wall will minimize these problems:

alternate courses

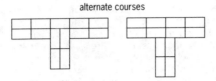

Fig. 5.7 8-inch joined brick T-intersection

alternate courses

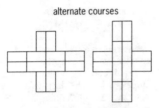

Fig. 5.8 8-inch brick continued intersection

The detail shown in [Fig. 5.9] is suggested as one method of building parapets. For a cavity wall construction, it is recommended that the cavity continue up into the parapet, thereby providing some flexibility between the outside wythe and the inner wythe. Expansion joints should extend up through the parapet. In addition, the parapet wall should be reinforced and doweled to the structural frame or have an additional expansion joint spaced between those in the wall below. Expansion joints should also be placed near corners to avoid displacement of the parapet. Parapet copings should provide a drip on both sides of

FRICTION FIT
METAL COPING

SEALANT

3/16" (4.8 mm) METAL TIES

HORIZONTAL REINFORCEMENT

VERTICAL REINFORCEMENT

2" (51 mm) MINIMUM CAVITY

COUNTER FLASHING

BASE FLASHING

DOVETAIL ANCHOR
SLOT AND
¼" (6.4 mm) FLEXIBLE
DOVETAIL ANCHOR @ 16"
(406 mm) O.C.
HORIZONTALLY

SOFT COMPRESSIBLE MATERIAL

PREFABRICATED
JOINT REINFORCEMENT

Fig. 5.9
Parapet wall

Brick Institute of America

the wall. Metal, stone, and fired clay copings of various design usually provide this feature. The back side of the parapet should be constructed of durable materials, preferably the same material that is used in the front side of the parapet. They should not be painted or coated, they must be left free to 'breathe.' Unless copings are impervious with watertight joints, place through flashings in the mortar bed immediately beneath them and firmly attach the coping to the wall below with anchor bolts.

(Courtesy Brick Institute of America)

RETAINING

Retaining walls are constructed to prevent hillsides from eroding or collapsing, to create terraces, and for other landscaping purposes. The four common types of retaining walls are gravity, cantilevered, counterfort, and buttressed. The last two types share many construction features and are considered as a single category in this discussion.

To increase resistance to water penetration, use full mortar joints in retaining walls. Use full-shell mortaring for concrete blocks. Select units and mortar to resist below-grade conditions. The wall must be fully set before backfilling. Allow at least five days for setting.

Retaining walls use the following terminology:

Toe: the portion of the footing on the downhill side of the wall.

Heel: the portion of the footing on the uphill side of the wall.

Key: a rectangular-section protrusion below the footing to prevent downhill slippage.

Wall plate or stem: the wall, exclusive of pilasters that support it.

EXCAVATION AND FOOTING

Observe rules for trench safety during excavation and construction. Trenches with slippery ground or a narrow working area present special hazards. Unshored trench walls may collapse during construction, especially if the soil is loose, traffic or rain is excessive,

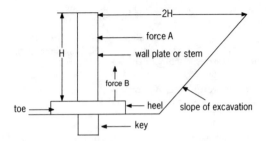

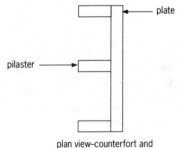

plan view–counterfort and
buttressed walls

Fig. 5.10 Parts of a retaining wall

or the walls are too steep. A good rule of thumb is to make the
sloping area twice as wide as the trench depth. If the ground is
uneven, smooth it or set up planks for walking during construction
(see *Safety, Trench*, p. 330).

Footings for retaining walls serve three purposes:

1. To support the weight of the wall. This is done by sizing the
   footing properly and placing it below the frost line, preparing a
   good excavation, and compacting loose soil if needed. Gravel is

permitted by some building codes to stabilize soil and increase drainage.

2. To resist sliding, or "creeping," down the slope. This is not a problem on shallow slopes in soil with a high degree of friction. In other cases, footings need a key to resist creep. The key is commonly made about as wide and thick as the slab.

3. To resist overturning. A gravity wall must be massive enough to resist pressure from the uphill earth. In cantilever and counterfort designs, the force of the uphill soil (force A in Fig. 5.10) tends to lift the heel of the footing (force B), and the weight of the soil prevents overturning. In the buttressed wall, the downslope force tends to push the toe into the earth.

Footings vary in design. Poured concrete can be either reinforced or unreinforced. If soil moisture is high, reinforcement is preferable. To key the plate to footing, place dowels in the footing where they can extend into a block core or the cavity between the wythes. An alternative is to form a keyway in the top of the footing.

Pour a footing by following these steps:

1. Locate the footing in the trench and mark off the exact location with stakes and lines.

2. Form the earth by cutting steep sides or placing wooden or steel forms. Compact soil that was disturbed during excavation, but do not fill voids with soil—use gravel instead.

3. Place horizontal reinforcement and dowels into position, and tie them to hold them steady.

4. Pour the concrete. When it is partially set, roughen the top surface to help mortar bonding.

5. Allow the footing to cure for the specified period before laying the wall.

WATER CONTROL IN RETAINING WALLS

A retaining wall that does not allow moisture to escape is likely to fail. Excess soil moisture can cause the following troubles in retaining walls: efflorescence, spalling, leaching, and frost heaving. The wall design should deal with water through a combination of weepholes, a drainage system, and water-resistant coatings. The need for such systems depends largely on weather and soil conditions at the site.

WEEPHOLES

Weepholes are set at intervals along the wall to allow water that accumulates to escape from the back. The weeps must slope about 1/2 inch per foot towards the exposed face; they are usually placed high enough to drain above the grade on the exposed side.

Weepholes can be constructed by removing sash cords from a filled head joint, by leaving a head joint dry, or by installing sections of steel pipe in a head joint. In block walls, you can make excellent weeps by setting hollow 8-inch × 8-inch × 8-inch units on their sides. (Use a second unit of equal size next to a hollow unit to maintain the bond.) Install steel mesh or screen in large weeps to prevent animals and insects from entering the wall.

During backfilling, shovel about one cubic foot of gravel near the weephole to prevent it from clogging with soil.

DRAINAGE

Gravel in combination with plastic drain tile is an excellent way of providing drainage at the base of a retaining wall. The tile must be located high enough so water on the downhill side cannot enter it. Slope the tile toward the outlets so water will not accumulate in it. Make sure the outlets drain freely.

Install drain tile in this sequence:

1. Lay up the wall. In counterfort and gravity walls, make provisions to install drains through the end pilasters or connect the tiles to a drain.

2. Apply waterproofing if required to the back of the wall.

3. Add gravel to the top of the footing on the uphill side, realizing that too much gravel will divert water away from the tile.

4. Lay out the drain tile, making sure it slopes from the center of the wall toward the outlets.

5. Terminate the drain tile at the wall ends. Cover the outlets with mesh to prevent clogging.

6. Add more gravel to bury the drain tile.

7. Begin backfilling. In dense soil, add gravel next to the wall to ensure drainage down the uphill side.

COATING

In areas with substantial soil moisture, parging or an application of bituminous sealer may be needed on the uphill side of the wall. Before waterproofing, tool the joints properly, both above and below grade. Depending on the nature of the masonry material and weather and soil conditions, two or three coats of parging or waterproofing may be required. Apply these materials using the same techniques used for basement walls (see *Coatings*, p. 195).

GRAVITY WALL

Gravity retaining walls depend on their weight and shape to withstand forces of the uphill earth. Compared to the other styles, they are simpler to construct and relatively less effective. Most gravity walls are limited to about 5 feet high. The base must be at least .7 times the height in thickness.

After excavation, follow these steps to construct a block gravity retaining wall, while observing blueprints and building codes as well (see Fig. 5.11):

1. Lay out and pour the footing as specified.

2. Center the wall on the footing and mark the wall and pilasters.

3. Determine the number of courses, the number of wythes per course, and the bonding pattern. Lay units dry if necessary.

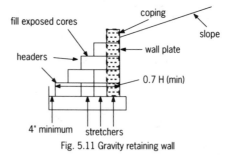

Fig. 5.11 Gravity retaining wall

4. Lay up the wall. Set lath under exposed cores of headers in the pilaster racking, then fill these cores with grout or mortar. Cut this filling flush with the top of the block.

5. Finish the wall with a coping course and caulk the joints.

6. Parge or coat the uphill side if necessary.

CANTILEVER WALL

Cantilever retaining walls transfer pressure from the uphill earth to the footing. This pressure tends to overturn the footing, but the heel is beneath the uphill earth and cannot lift. Construction techniques for reinforced masonry, low-lift grouting, and cavity walls are all used for cantilever walls.

Some cantilever walls are battered, or sloped, on the exposed face. These walls are slightly wider at the bottom than at the top, so that the wall will have a vertical face when it settles into place. Battering can be expressed in two ways in blueprints: 1) In inches per vertical foot; a slope of 1/2 inch per foot means that the wall departs from vertical by 1/2 inch per foot of height. A wall 4 feet high with a slope of 1/2 inch per foot tapers 2 inches over its height. 2) Thickness of first course and thickness of last course. Usually, the uphill wall is made vertical and the facing wall given the angle, so make the rear wythe vertical and measure from it.

After reading plans and specifications, lay a cantilevered wall as follows:

1. Pour footings and lay in dowels at specified locations. Roughen the top of the footing or provide a keyway to increase the resistance to sliding.

2. Set in drain tile if needed.

3. Mark the location of the wall and pilasters. Lay out the first course dry to check the bond. Space the units so the dowels are in the center of a unit core or the cavity between wythes.

4. Lay the first course in a full bed of mortar, forming cleanout holes for the grouted cores. Form these holes by leaving out a unit or cutting off face side of a unit.

5. Lay the rest of the wall.

6. Fill the cores with low-lift grouting procedures (see *Grouting, low lift,* p. 242).

7. Install coping and coat the rear face if needed.

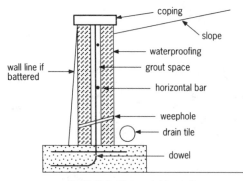

Fig. 5.12 Cantilever retaining wall

COUNTERFORT AND BUTTRESSED WALLS

Both counterfort and buttressed walls depend on diagonal struts in the pilaster to transfer the force of the earth to the footing. In counterfort walls, the strut is in tension; in buttressed walls, it is in compression. Pilasters are visible on buttressed walls and invisible on counterfort walls.

Follow these steps to build a counterfort or buttressed retaining wall:

1. Pour the footing in proper location. The lower ends of struts for counterfort walls have a hook shape. Insert the struts in the footing at the proper angle, then brace the upper ends and pad them for safety. Roughen the footing top as it hardens to increase the bond between the footing and the wall.

2. Locate the plate along the footing, then mark the pilaster locations.

3. Lay the first course dry, and determine the bonding between pilasters and the plate.
4. Begin laying the wall, placing the pilaster units around the reinforcing rods. Install joint reinforcing if specified.
5. Grout pilaster cavities after the wall is completed, cutting off the tops of cores flush.
6. Install the coping. Install drain tiles if necessary. Coat the wall as needed.
7. Backfill after the wall has cured.

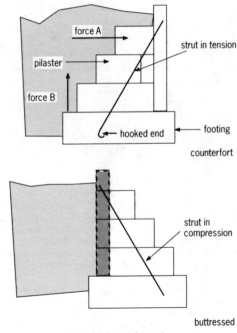

Fig. 5.13 Counterfort and buttressed retaining walls

BACKFILLING

Backfilling must wait until the wall has gained sufficient strength—usually one week or more. In cold weather, allow longer curing time. If drain tiles are to be installed, they must be embedded in gravel next to the footing before backfilling. Some applications, especially in heavy clay soils, require extra gravel near the wall to improve drainage.

During backfilling, observe these precautions:

- *Make sure all waterproofing operations have been finished.*
- *Keep heavy equipment at least one wall height away from the wall.*
- *Temporarily brace the wall to prevent heavy equipment from upsetting it.*
- *Fill areas around weepholes with gravel to prevent clogging with soil.*
- *Break up large chunks of soil before placing them near the wall.*
- *Remove large stones from the backfill.*
- *Compact the backfill after every 2 feet of filling.*

SCREEN

Screen walls reduce sunlight and wind while allowing diffuse light and ventilation to pass through. Screen walls are usually constructed of special screen blocks, although hollow 8-inch blocks turned sideways will make a screen wall.

Type S or M mortar is used for exterior walls; Types S, M, or N can be used for interior walls. Use full mortar joints to reduce water damage in exterior walls. If any hollow units are laid with vertical voids exposed to the weather, provide a coping or capping to prevent water entry.

Nonbearing screen walls need a minimum nominal thickness of 4 inches and a clear span no more than 36 times the nominal thickness (12 feet for a 4-inch wall). Loadbearing walls should have a nominal thickness of 6 inches.

Although screen walls seldom carry loads (and are prohibited from doing so by some building codes), they must be able to resist wind and other lateral loads. Four techniques can be used to ensure stability:

1. Use a framing system that transfers wind loads to the ground.
2. Connect the wall to another framing system.
3. Limit the length of unsupported wall.
4. Use vertical and horizontal reinforcement.

Use techniques of reinforced masonry to construct a large concrete block screen wall. Pilasters should have vertical reinforcement. Anchor the horizontal reinforcement to the pilasters. Use bond beams at the top of wall panels.

Screen walls can use piers, intersecting walls, columns, or buttresses for lateral support. Structural steel, including channels and flat bars, can be used for vertical reinforcing. A screen wall can be tied into buildings with anchors and reinforcing rods.

SERPENTINE

Serpentine walls are among the most graceful masonry structures. Their distinctive S shape gives them not only great beauty but also great strength, allowing unsupported 4-inch walls far longer than those built with other techniques. Though serpentine walls are complicated to build, they are economical with materials.

Solid units are preferred. Grade SW bricks should be used unless the weather is very mild. Tool carefully to avoid moisture problems. Choose the coping to conform to the curvature and to complement the appearance of the wall.

READING BLUEPRINTS

Serpentine walls have their own terminology (see Fig. 5.14):

Centerline: The line from end to end which divides the wall in half.

Radius point: The point at the center of a bay used to mark that bay. A radius point is not necessarily on the centerline.

Radius line: The line connecting a radius point to the arc in the same bay. Used to mark out the arcs for each bay.

Arc: The curving line of the wall.

Bay: The area described by an arc between two crossings of the centerline.

Tight arc: A small-radius arc at a wall end. Used to support the end of the wall instead of a pilaster.

Depth of bay: Distance from the centerline to the furthest point of the arc in that bay. Not always equal to the radius. Double the depth to find the approximate width of land required.

Span: The width of a bay measured along the centerline.

Length: The distance along the centerline from end to end. This is shorter than the actual wall length but it provides a measurement of how much land a wall requires.

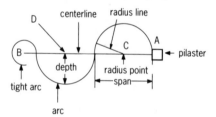

Fig. 5.14 Serpentine wall elements and layout

LAYOUT

Use the following technique for serpentine walls with semicircular bays. In this design, radius points are located on the centerline and the depth of the bay equals the radius. Mark all points with stakes.

1. Consult drawings to find span, radius, and the location for the wall.

2. Determine the location for the centerline and drive stakes (A) and (B). Then snap a line or stretch the centerline (AB).

3. Mark the wall termination—a pilaster, an attachment to another structure, or a tight arc.

4. Mark the radius points along the centerline. Start from the intersection of the wall and the pilaster, (A). Measure one-half span from (A) along the centerline to mark (C), the first radius point. Then measure one full span from (C) and mark (D), the second radius point. Mark all remaining radius points in the same manner. Then remeasure along the centerline and hammer a nail into each stake at the exact radius point.

5. Make a radius line. Loop one end of a line on a nail on the radius stake and stretch it toward the centerline. Attach a pencil to the line at the exact length of the radius.

6. Mark the first bay by swinging the pencil on the radius line. The marking technique will depend on the surface. For rough dirt or grass, spread sand or chalk first and mark the arc with a pencil or stick held on the line. On concrete, use a heavy-duty pencil or crayon.

7. Mark the other bays with the same line, making sure that the arcs make a smooth curve as they pass the centerline.

8. These lines just marked are the center of the wall. Mark the edge of the wall by measuring 1/2 the wall thickness to one side of the lines just laid out.

9. Check the layout visually by laying a dry course and checking it by eye. Then measure again to ensure everything matches the plans.

A different method must be used for walls with a more gentle curve. In these walls, radius points are located on a line perpendicular to the centerline, not on the centerline. The depth of the bay is less than the radius. Mark all points with stakes during layout (see Fig. 5.15).

1. Consult drawings to find span, radius, and wall location.

2. Insert stakes at (A) and (B), the ends of the centerline. Draw the centerline (AB).

3. Measure one-half span from (A) and mark (C), the center of the first bay.

4. Mark (D), the end of the first bay by measuring one full span from (A).

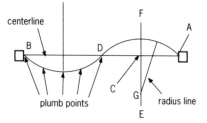

Fig. 5.15 Serpentine wall with shallow bays

5. Draw line (EF) perpendicular to the centerline through (C).

6. Find the radius point for the first bay. Measure one radius from (A) and (D) to line (EF). Both intersections must fall at the same spot on (EF). Mark the radius point (G).

7. Make a radius line. Attach one end of a line to a nail on stake (G) and stretch the line to the radius, length (DG). Tie a pencil to the line at the exact radius length.

8. Draw arc (AFD) along the far side of centerline.

9. Repeat this procedure for each remaining bay.

10. Mark the edge of the wall by measuring one-half the wall thickness to one side of the arcs just laid out.

11. Check the layout by eye and by measuring distances of the various dimensions.

LAYING THE WALL

1. Lay out the wall dry, checking the bond and intersections with pilasters and tight arcs. Remember that the first line marked above was the center of the wythe, not the edge of the wall. Head joints should taper to no thinner than 1/4 inch in the arcs. Mark the head joints on the foundation.

2. Locate plumb points along the wall, about 2 to 3 feet apart. Each bay usually has five plumb points: one at each end, one halfway between the ends, and one half-way between this point and each end. Mark these points clearly on the foundation and the template.

3. Cut a piece of 3/4-inch plywood as a template. Lay the plywood on the wall location and mark it using the radius point and radius line used for the wall layout. The template must extend at least 18 inches beyond adjacent plumb points. Mark plumb points on the template and carefully saw it to shape.

4. Begin the first course by placing units at each plumb point. Allow these units to bond briefly and place the template alongside the arc. Always match the plumb point marks on the template with the plumb points on the wall to ensure that you always place the template in the same position.

5. While one bricklayer holds the template, another lays the units between the plumb points.

COPING THE WALL

After the final course has been laid, install a flashing and coping:

1. Cut flashing (often roll roofing) 1 inch narrower than the thickness of the wall.

2. Spread roofing cement on the top of the wall, leaving about 1/2 inch at the outside edge free of cement.

3. Lay the flashing onto the cement, cutting it to conform to the curve.

4. Add extra roofing cement over the cuts in the flashing.

5. Lay out a bed of mortar and install the coping.

VENEER

Veneer walls are nonbearing wythes attached to structural elements of a building. A common application is on houses, in which case the veneer is secured to the framing with wall ties. Building codes govern placement of wall ties. They are often placed at 16 inches

O.C. vertically and 32 inches O.C. horizontally. Place extra ties around openings and other stress points (see Fig. 5.16).

In many ways, veneer walls are similar to cavity walls. The air space provides some insulation and helps prevent water from entering the building. Flashing and weepholes at the bottom prevent water that enters from remaining in the cavity (see *Walls, cavity*, p. 94).

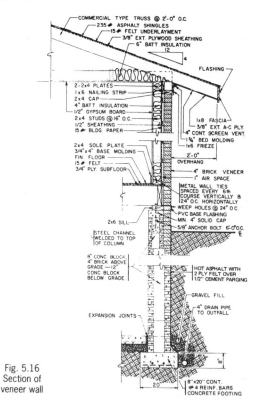

COMMERCIAL TYPE TRUSS @ 2'-0" O.C.
235 # ASPHALT SHINGLES
15 # FELT UNDERLAYMENT
3/8" EXT. PLYWOOD SHEATHING
6" BATT INSULATION

FLASHING

2-2x4 PLATES
1x6 NAILING STRIP
2x4 CAP
4" BATT INSULATION
1/2" GYPSUM BOARD
2x4 STUDS @ 16" O.C.
1/2" SHEATHING
15 # BLDG. PAPER

1x8 FASCIA A-C
3/8" EXT. A-C PLY.
4" CONT. SCREEN VENT
1¾" BED MOLDING
1x6 FRIEZE

2x4 SOLE PLATE
3/4"x 4" BASE MOLDING
FIN. FLOOR
15 # FELT
3/4" PLY. SUBFLOOR

2'-0"
OVERHANG

4" BRICK VENEER
1" AIR SPACE

METAL WALL TIES
SPACED EVERY 6th
COURSE VERTICALLY &
24" O.C. HORIZONTALLY
WEEP HOLES @ 24" O.C.
PVC BASE FLASHING
MIN. 4" SOLID CAP
5/8" ANCHOR BOLT 6'-0" O.C.

2x6 SILL

STEEL CHANNEL
WELDED TO TOP
OF COLUMN

8" CONC. BLOCK.
4" BRICK ABOVE
GRADE — 12"
CONC. BLOCK
BELOW GRADE

HOT ASPHALT WITH
2 PLY FELT OVER
1/2" CEMENT PARGING

GRAVEL FILL
4" DRAIN PIPE
TO OUTFALL

EXPANSION JOINTS

8"

20"
8"x20" CONT.
4 REINF. BARS
CONCRETE FOOTING

Fig. 5.16
Section of
veneer wall

STEEL STUD VENEER WALL

The practice of using steel studs to back veneer walls allows for insulation and enables efficient scheduling of construction. Because steel and masonry have different rates of thermal expansion, this type of wall must allow for differential movement. When building such a wall, consider the following:

1. The proper selection and spacing of ties.
2. Details to prevent water entry: flashing, weepholes, joint sealant, and air spaces.
3. Strength and stiffness of the wall.
4. Corrosion resistance of all metal components.
5. Climatic and interior conditions.
6. The need for horizontal and vertical movement joints.

THIN VENEER

Thin veneer is a type of brick with various face dimensions and a thickness ranging from 1/2 inch to 1 inch. The units can be adhered to a surface or panels can be prefabricated on the ground.

The thin-set method uses an epoxy or organic adhesives to bond the veneer units to a backing. The technique is suitable for interior work only. In damp areas, the backing must be clean and dry masonry or concrete. In dry locations, wood and wallboard are acceptable backings.

The thick-set method is permissible for interior and exterior use over a backing of masonry, concrete, steel, or wood. A layer of wire lath can be fastened to masonry or concrete to hold the scratch coat. A moisture barrier should be used over wood or steel backing. Apply a scratch coat and a bond coat over the backing, then press the veneer units into the bond coat.

Thin brick units can also be prefabricated with a casting technique. Prefabricated panels may be cast into complex shapes. Because they are relatively light, in some cases there is no need for scaffolding during installation. To cast, the units are placed face down in a form and a fluid grout is flowed over the back and into the voids between units. After the grout sets, a backing of concrete and reinforcement is cast. The panel is allowed to cure then placed in position.

PART THREE

FIREPLACE AND CHIMNEY

6
Fireplace

Fireplaces are a big selling point in homes; they have always provided radiant energy and a feeling of warmth to people sitting directly in front of them. But conventional fireplaces can remove so much heated air from the home that they become wasters, not sources, of energy. Two solutions to this problem are in common use: glass doors and heat-circulating fireplaces. Glass doors permit only as much air as is needed for combustion to exit the building through the chimney. They also prevent sparks from escaping the firebox and reduce smoking and downdrafts. Heat-circulating fireplaces are devices that channel room air behind the firebox— where it is heated—then back to the room. These units, which are usually prefabricated steel, are very popular because they increase efficiency and reduce labor requirements. Often, circulating fireplaces are used in conjunction with glass doors.

Because so many types of fireplaces are possible, exact plans cannot be given here. Fireplaces are constructed with certain relationships among the various key dimensions. Guidelines, such as Table 6.1, and experience are both essential to building proper fireplaces.

RUMFORD FIREPLACE

The Rumford fireplace was designed centuries ago to increase the radiant heat output by building a shallow unit with flaring sides. The design has historic importance and may be needed in restoration work, but keep in mind that a proper Rumford fireplace might violate building codes, and the units are prone to smoking when

TABLE 6.1
Conventional Fireplace Dimensions

A	B	C	D	E	F	G
24	24	16	11	14	19	8 × 12
26	24	16	13	14	21	8 × 12
28	24	16	15	14	21	8 × 12
30	29	16	17	14	24	12 × 12
32	29	16	19	14	24	12 × 12
36	29	16	23	14	27	12 × 12
40	29	16	27	14	29	12 × 16
42	32	16	29	14	32	16 × 16
48	32	18	33	14	37	16 × 16
54	37	20	37	16	45	16 × 16
60	37	22	42	16	45	16 × 20
60	40	22	42	16	45	16 × 20
72	40	22	54	16	56	20 × 20
84	40	24	64	20	61	20 × 24
96	40	24	76	20	75	20 × 24

COURTESY OF THE BRICK INSTITUTE OF AMERICA

A   WIDTH OF OPENING
B   HEIGHT OF OPENING
C   DEPTH OF OPENING (MEASURED TO REAR OF FACE)
D   REAR WIDTH OF FIREBOX
E   REAR HEIGHT OF FIREBOX FROM FLOOR TO START OF SLOPE
F   HEIGHT OF SMOKE CHAMBER
G   INSIDE DIMENSIONS OF FLUE

cold. A flat-plate damper should be used; if this is unavailable, a chimney-top damper can be installed (spring-loaded to remain open if the operating mechanism fails).

BUILDING CODE REQUIREMENTS

Fireplaces must, by code, have minimum clearances to combustible materials. These gaps are sometimes filled with firestops, which are sheet-metal linings that prevent sparks that have penetrated cracks in the masonry from igniting the structure. Firestops may also be filled with other heat-resistant, noncombustible materials, such as fiberglass.

These organizations provide standards and information related to fireplace safety:

National Fire Protection Association. NFPA 211, Standard for Chimneys, Fireplaces, and Vents.

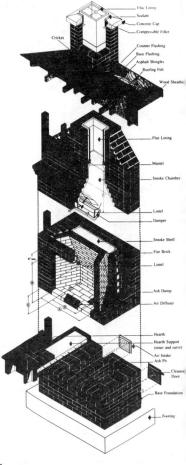

Thumb Rules:

1) The height of the fireplace opening, Ⓑ, should be at least 6″ less than the width of the fireplace opening, Ⓐ.
2) The back width of the firebox, Ⓓ, should be 1′ less than the front width of the fireplace, Ⓐ.
3) The back height of the firebox, Ⓔ, should be no more than ½ the opening height of the fireplace, Ⓑ.
4) The height of the smoke chamber, Ⓕ, should be about equal to the height of the fireplace opening, Ⓑ.

Before building a fireplace and chimney, check with the local building codes.

Fig. 6.1
Fireplace details

Courtesy of Brick Institute of America

American Insurance Association. Code for the Installation of Heat-Producing Appliances and Heating, Ventilating, Air Conditioning, and Blower and Exhaust Systems.

Check applicable building codes for details on fireplace construction, especially pertaining to:

- *Hearth size.*
- *Construction of the firebox.*
- *Flue (must usually be lined and have 4 inches of masonry around the liner).*
- *Clearance between chimney and any wood building elements. (In some codes, clearance may be reduced by filling with firestops, a noncombustible material such as steel, fiberglass, or mineral wool.)*
- *Placement of mantel and other combustible elements near the opening. (In general, do not place combustible elements within 6 inches of the opening.)*

DRAFT

Draft is another name for convection, the force which draws smoke from the fireplace. Convection is the physical force by which denser fluids push aside less dense ones. In fireplaces, convection occurs because smoke is hotter and less dense than room air. The hotter the smoke, all other things being equal, the greater the convection and draft, and the greater the probability that the fireplace will exhaust properly. As heat is extracted from the smoke, the smoke cools. Thus, the desire to heat the room conflicts with the need to produce a draft that is strong enough to remove smoke.

Cool smoke is likely to cause fireplace smoking problems. Another cause of smoking is chimneys on outside walls, which operate at lower temperature, thus cooling the smoke, and drawing more poorly than warmer, interior chimneys.

The inside area of the flue must be about one-tenth the area of the fireplace opening. A large flue will draw better than a small one, but if the flue is too large, it will not heat up, and the draft will be reduced (and the chimney will be expensive to build). When closed, glass doors greatly reduce the opening of a fireplace. However, the flue size cannot be decreased because the chimney must still draw when the doors are open.

LAYING FIREBRICK

Firebrick, a blend of special clays that resist heat, is generally used to build the firebox. Firebrick (also called refractory brick) should be mortared with refractory cement, which can also withstand heat, although conventional mortar is sometimes used. Refractory cement may be purchased dry or premixed.

Firebrick must be completely dry before laying. Dip the brick into a soupy mix of refractory cement, place the brick, then tap it to seat thoroughly, ensuring a tight fit with full joints. Allow the cement to set for thirty days before using the fireplace.

PARTS

Fireplaces have unique parts, and a jargon all their own. These parts are described below:

ASH DUMP

An ash dump allows the fire tender to transfer ashes directly from the firebox to the soot pocket for eventual removal through the cleanout door. The ash dump is a great convenience in keeping the room clean. Dumps are usually made of cast iron and the same size as a firebrick so they can be installed easily in the inner hearth. Build an opening in the structural hearth beneath the ash dump to allow ash to fall into the soot pocket. Access to the soot pocket is provided by a cleanout. Ash dumps are sometimes combined with external air inlets.

CHIMNEY

Chimneys have two related functions—to provide draft, and to dispose of smoke. Chimneys for woodburning devices, including fireplaces, should have a spark arrestor, preferably of noncorrosive material, at the top. Poor chimney design and construction are responsible for many fireplace hazards and problems (see *Chapter 7, Chimney*).

CLEANOUT

A door should be provided at the base of each flue to clean out accumulated debris. In most fireplaces, the soot collects at the smoke shelf, so the cleanout is used only to remove ash from the ash dump. The cleanout door can be in the basement or on the outside of the chimney.

DAMPER

A damper is a steel or cast-iron flap that seals the throat when the fireplace is not in use. The purpose is to prevent warm air from escaping in winter or entering in summer. A damper must swing freely when operated, provide a good seal when closed, and not impede the passage of smoke when open.

Dampers are controlled by a rod, poker, or chain. Cast-iron dampers are preferable because of their simplicity, tolerance to heat, and durability. "High-formed" dampers are useful because they extend the smoke chamber and increase the draft.

The damper sits toward the front of the smoke chamber to allow room for a smoke shelf behind it. Because the damper is metal, you must allow room for expansion when installing. Seat the damper on a thin layer of refractory cement, but do not embed it. Never allow any masonry units to bear on or against the damper. Fill the area between the damper and adjacent masonry with noncombustible, compressible insulation.

EXTERNAL AIR INLET

There are two reasons to provide an external air intake for combustion air: 1) Most new homes are relatively airtight, and fireplaces may not be able to draw enough air to burn or dispose of combustion product effectively. 2) Fireplaces waste energy by removing heated air from the building, and this loss is reduced if external air is burned. Many building codes require provisions for outside combustion air. The inlet may be combined with the ash dump.

The inlet should be built entirely inside the masonry chimney because it can become a secondary flue. In some conditions, fire can spread down the air intake and ignite other building elements. Tight dampers on the inlet will prevent infiltration of air when the fireplace is not being used and also will control air intake to prevent flooding the combustion chamber with cold air. The inlet size should be proportional to the opening size. The inlet should not be located at the back of the firebox, as this can blow ashes toward the opening. Construct a chamber just below the hearth to slow down the intake air. This will prevent hot spots in the firebox, which can damage andirons, intake grates, and glass doors.

FACE

The face is the visible portion of the fireplace. It may be built of any masonry material—with brick, tile, and stone being the most common. In some applications, the face is plaster or wallboard and the only visible masonry is the columns of brick surrounding the opening.

FIREBOX

The firebox is where the fire is built and maintained. It can be built of firebrick (at least 2 inches thick and mortared with refractory cement) or prefabricated steel. Most codes require 8-inch solid masonry around the firebox (firebox wall plus backing wall) and 2-inch clearance between fireplace or chimney, and all combustible materials.

Firebox design is essential to proper operation. The walls should slope inward (called "splaying") to reflect heat and channel the smoke toward the throat. See Table 6.1 for common firebox proportions.

The firebox is usually constructed on a concrete slab. The backing wall, usually concrete block or brick, is brought to a height of about 5 feet before the firebrick is laid. The backing wall should be at least 4 inches thick to support the weight of the throat and chimney. Some masons fill the gap between the firebrick and backing material with mortar and brick scraps, but it is best to allow some room for expansion of the firebox. This can be done by filling the gap with fiberglass or other compressible insulation. The firebox should cure for thirty days before a fire is laid.

FLUE

Fireplace chimneys must be lined with flue liner to 1) provide a "first line of defense" against smoke and sparks leaving the chimney, 2) to prevent damage from chimney fires, and 3) allow easy removal of condensed smoke (creosote). The flue liner rests on the top of the smoke chamber (which is corbelled in to bear the weight or has a lintel for the same purpose). The top surface of the final flue should be horizontal to reduce the chance of downdrafts. Multiple flues should terminate at different heights to prevent smoke from one flue from entering another and returning to the building (see *Chimney*, p. 141).

FOOTINGS

Footings must be large and strong enough to support the fireplace and chimney without cracking or shifting. Minimum standards generally are:

- *Concrete at least 12 inches thick, extending at least 6 inches beyond the fireplace in all directions.*
- *Set below frost line unless located in an area not subject to freezing.*
- *Placed on properly-prepared soil (undisturbed or filled with compacted sand or gravel).*
- *Foundation walls (between the footing and the fireplace) should be masonry or concrete, at least 8 inches thick.*

GLASS DOORS

Glass doors change the operation of a fireplace by 1) reducing the amount of air entering the fire and leaving the room, 2) channelling air across the fire to fan the flames, and 3) reducing the amount of radiant heat given to the room. Glass doors are incompatible with some manufactured fireplaces; consult the manufacturer. Glass doors must be protected from cracking due to heat, and from soiling due to close encounters with smoke. The door assembly should close tightly and be sealed around the perimeter to prevent energy loss when the fireplace is not in use.

Follow these practices when using a fireplace with glass doors:

- *Use substantial grates or andirons to keep the fire 6 inches to 8 inches from the doors.*
- *Do not burn paper, plastic, or trash. Use no more kindling than necessary.*
- *Start the fire slowly to allow the doors to warm up evenly, and avoid quick, hot fires.*
- *Use draft controls on the doors to maintain a medium-size fire.*

- *If the fire becomes large enough to lick at the doors, open them wide so the flames cannot reach them. Keep the doors open until the fire dies down.*

- *Do not operate the fireplace with doors partly open.*

HEARTH

Hearths have two sections—inner and outer. Both are supported by reinforced masonry or, more commonly, a reinforced slab. Make sure to allow for any ash dump and external air inlet when building the hearth support. The inner hearth, the base of the firebox, is usually built of firebrick.

The outer hearth, the fireproof area in front of the firebox, may be built of any flat masonry units, such as block, brick, pavers, or tile. For small openings, the hearth should extend 8 inches to the sides of the opening and 16 inches in front. For larger openings (above about 6 square feet) the side dimension should be 12 inches and the front dimension 20 inches. See local building codes for minimum hearth dimensions.

The outer hearth is an important element of a fireplace's visual impact. Raised hearths are popular because they bring the fire closer to eye level, make it easier to tend, and provide a seating area. Raised hearths can be extended to the sides to increase the seating.

LINTELS

The lintel above the opening supports the smoke chamber and part of the chimney. A second lintel is required over the damper to support the smoke chamber and flue. The opening lintel must seat securely on the face. Lintels can be made of stone, steel or reinforced concrete. The ends of steel lintels should be wrapped with fibrous, fireproof insulation to allow expansion.

For the opening, you can use a masonry arch instead of a steel beam. Use a center to lay the arch and support it while the mortar sets (see *Arches*, p. 158).

MANUFACTURED FIREPLACE

Steel fireplaces have been controversial in the masonry trade. Many masons feel they substitute a structure that will eventually rust out for a quality, durable masonry fireplace. But the reality is that manufactured fireplaces have significant advantages in terms of heating and installation costs. These units usually use fans to circulate air behind the fire and return it to the room. Because many homeowners demand these units, masons must learn to choose and install them.

Many styles of manufactured fireplaces are available; options include prefinished or masonry fronts, one-, two-, or three-sided openings, and various blower and air-handling arrangements. Ash dumps, hookups to furnace air ducts, and log lighters are also available. Some devices specify a masonry or a metal chimney, in other cases the builder can choose either type. Manufacturers provide detailed installation instructions.

Courtesy Superior, The Fireplace Company

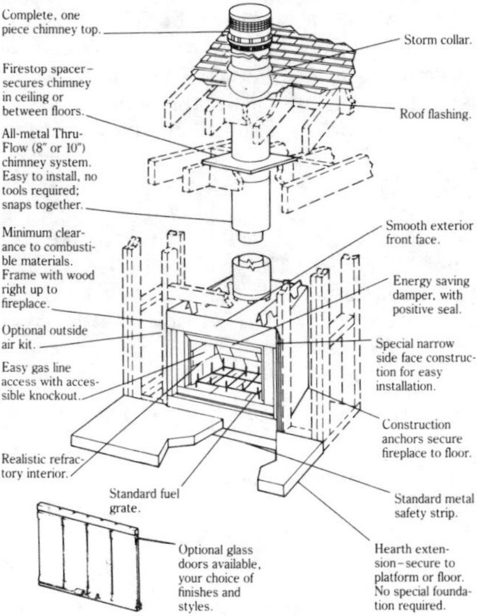

Complete, one piece chimney top.

Firestop spacer – secures chimney in ceiling or between floors.

All-metal Thru-Flow (8" or 10") chimney system. Easy to install, no tools required; snaps together.

Minimum clearance to combustible materials. Frame with wood right up to fireplace.

Optional outside air kit.

Easy gas line access with accessible knockout.

Realistic refractory interior.

Standard fuel grate.

Optional glass doors available, your choice of finishes and styles.

Storm collar.

Roof flashing.

Smooth exterior front face.

Energy saving damper, with positive seal.

Special narrow side face construction for easy installation.

Construction anchors secure fireplace to floor.

Standard metal safety strip.

Hearth extension – secure to platform or floor. No special foundation required.

Fig. 6.2
Fireplace Insert

OPENING

The width and height of the opening are the basic dimensions used to designate fireplace size. For example, a fireplace might be described as a 24 inches by 30 inches, in which case the height of the opening would be 24 inches and the width 30 inches. The height is usually about two-thirds of the width. The opening size, in turn, helps determine damper and flue size. See Table 6.1 for conventional dimensions of fireplaces.

SMOKE CHAMBER AND SHELF

The smoke chamber is above the throat, at the bottom of the flue. The chamber has two essential roles: to compress the smoke and channel it to the chimney, and to catch downdrafts. The smoke shelf, which forms the bottom of the chamber, has a curved bottom to deflect downdrafts back up the chimney and prevent them from entering the firebox.

The rear of the chamber should be vertical, and the other three sides should slope inwards to meet the flue. Metal lining plates are available to simplify smoke chamber construction. If the masonry is less than 8 inches thick at this point, the chamber should be parged with a 3/4-inch layer of refractory mortar for fire protection.

THROAT

Smoke passes from the firebox through the throat on its way to the smoke chamber. The size and design of the throat are critical for proper operation. The throat should be at least 8 inches above the front of the opening. The damper is installed in the throat to seal the chimney when the fireplace is not in operation.

EQUIPMENT CHECKLIST

The following list includes most common materials required for fireplace construction. Use this as a checklist when preparing estimates, ordering materials, and loading the truck:

Information:

Other appliances and devices using same chimney

Size, location, and type
Other contractor's tasks

Equipment:

Scaffold and hoist
Mixer

Footing and hearth forms

Materials:

Joint reinforcement
Damper
Ash dump
Fans and wiring
Barbecue unit
Concrete and form for
    hearth
Firebrick and refractory
    cement
Flue lining
Mantle material and
    anchors for wood
Flashing
Caulking
Cleanup chemicals

Cleanout door
Lintels
Manufactured fireplace
Glass doors
Insulation
Hearth stones, tiles, or
    pavers
Bricks and mortar
Block for backup walls
Weather protection or
    spark arrestor for flue
Chimney cap or forms for
    casting
Sealant

PLANNING A FIREPLACE

The first step in building a fireplace is to decide upon its size, location, and type. The Brick Institute of America recommends certain relationships be used in key fireplace dimensions (see Table 6.1 for rules of thumb on this). With a manufactured fireplace, these relationships are determined by the manufacturer.

The size, shape, and construction details should all be chosen to harmonize with the location and provide a safe, durable structure. When planning a fireplace, remember that masonry must only rest on other masonry, as building on wood is unstable, illegal, and prone to fire damage.

If the location is not specified, or the homeowner is unsure, take time to discuss the alternatives. Have the homeowner answer the following questions about fireplace type:

- *Will it have one or more openings?*

- *How big should it be?*

- *Will it be all masonry or use a steel firebox?*

- *What type of face and mantle are desired?*

Now have the homeowner answer these questions about location:

- *Is a furnace thermostat nearby? (The fireplace may throw off the thermostat, although the thermostat can be moved if this is the only problem with a location.)*
- *Is a hot air outlet nearby? (The vent may blow ashes or disturb the flames, although it can be shut off during fireplace operation.)*
- *Is the fireplace out of the flow of foot traffic? Does the family have enough area to gather in front?*
- *Is the location near a stairway to the upstairs? (This can funnel heat upstairs and prevent the room with the fireplace from warming up.)*
- *Is outside air available for an external air inlet? Will you be able to install an ash dump?*
- *Must other appliances or fireplaces use the same chimney? (Use of the same flue is prohibited by many codes.)*
- *Can you observe necessary clearances between the opening and combustible surfaces? (Check with building inspectors for exact clearances.)*
- *If the fireplace is being installed in an existing building, how will it fit with the structure? Is there room to build the footing? How will the chimney fit with the roof line?*
- *Have you checked that no trees or other buildings overhang the chimney? (These can cause downdrafts, although trees can be trimmed if needed.)*

SAFE OPERATION

A mason must take time to ensure that a customer understands these basics of fireplace operation:

- *Keep the fire back from glass doors (see Glass doors, p. 135).*
- *Burn cured wood only.*
- *Do not use coal except in fireplaces designed for it.*
- *Avoid slow, smoky fires.*
- *Clean the chimney regularly.*
- *Burn proprietary chimney cleaners in the fire as directed.*
- *Keep alert for signs of a chimney fire and call the fire department if one is suspected. Inspect a chimney after a fire.*

See also *Chimney; Troubleshooting*, p. 151.

7
Chimney

Because chimneys exhaust the byproducts of combustion, safety is the primary consideration in planning and constructing. Interior chimneys for woodburning appliances must have 2-inch clearance to combustible materials. This gap can be filled with noncombustible material such as fiberglass. In exterior chimneys, a 1-inch clearance is required. The air space can be bridged by nonflammable siding or sealed with sealant.

A chimney can house one or more flues of similar or different sizes. Most building codes require that individual flues be separated by a 4-inch nominal masonry wall. Flues should have clearance so they can expand when they warm up. Chimneys and flues should not change size or shape within 6 inches of floor, ceiling, or rafters.

A chimney must be tall enough to provide a good draft in all weather conditions. Most codes require at least 3 feet of projection above the roof at the spot where the chimney penetrates the roof. On peaked roofs, the chimney must project 2 feet above any part of the roof within 10 feet horizontally of it. Chimneys that draw poorly can often be improved by extension.

Exterior chimneys must be anchored securely to the framing (see below). Check local codes for requirements on chimney construction.

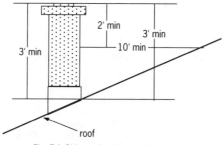
Fig. 7.1 Chimney height guidelines

CATEGORIES OF CHIMNEY

RESIDENTIAL

Residential chimneys are used for fireplaces, furnaces, water heaters, stoves, and other small fuel-burning appliances. These chimneys are also called "Class A" chimneys, and are used for any solid fuel and for gas, oil, or propane. Outlet gases entering residential chimneys should be cooler than 800°F. Residential chimneys are sometimes fully described on blueprints; in other cases, their location may be specified and the construction details left to the mason.

COMMERCIAL

Commercial chimneys exhaust gases from bakery ovens, furnaces, steam boilers, and wood or gas retorts. Commercial chimneys are used on factories, schools, apartments, shopping centers, and hospitals; they are generally designed to handle gases up to 1000°F. Some commercial establishments may use residential-style chimneys; check local codes.

Commercial chimneys are termed low- or medium-heat industrial chimneys in some codes and standards. Building codes often require that the flue liner in commercial chimneys be surrounded by an 8-inch masonry wall.

INDUSTRIAL

Industrial chimneys may designed to handle temperatures ranging beyond 2000°F. These chimneys are connected to blast furnaces, kilns, various types of ovens, and other sources of extreme heat. Industrial chimneys are designed by specialists.

REINFORCING AND ANCHORING

Chimneys are vulnerable to wind and earthquakes. A special set of specifications has been developed for chimneys in areas where earthquakes are possible. These standards require that chimneys be built of reinforced masonry construction. Reinforcing may also be needed for especially tall chimneys or those subject to high winds.

To reinforce a chimney, insert reinforcing rods into the cavity between the flue and the chimney wall after you have laid up a section of chimney. Then pour grout or mortar into the cavity (see *Grouting*, p. 240). Also insert 1/4-inch rods bent to the proper shape in bed joints 18 inches O.C. Consult building codes for details (see Fig. 7.2).

Exterior chimneys must be anchored to framing during construction at floor, ceiling, and roof levels. Bolt these anchors securely to framing members and insert the anchor's tongue into a head joint; then lay the course (see Fig. 7.3).

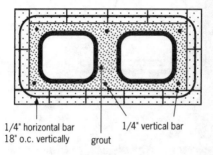

1/4" horizontal bar
18" o.c. vertically   1/4" vertical bar
                  grout

Fig. 7.2 Reinforced chimney

WATER RESISTANCE

Flashing must make a raintight seal between the roof and the chimney protruding through it (see *Flashing*, p. 231). In addition, you must make a watertight seal between the chimney and the building siding by caulking the gap. The top of the flue may be fitted with a cap to prevent water from running down the flue.

PARTS OF A CHIMNEY

CAP

Precast chimney caps are available for common sizes of flue and chimney. Other sizes can be custom-cast. Small caps can be precast on the ground. Large caps, due to their weight, should be cast in place.

Reinforcing mesh or rod should be cast into the cap. Slope the cap away from the flue for drainage. Provide a drip edge at the outside

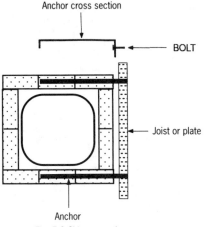

Anchor cross section

BOLT

Joist or plate

Anchor

Fig. 7.3 Chimney anchor

bottom to prevent water from dripping down the chimney (see Fig. 7.4).

If multiple flues are used in a single chimney, extend them to different heights to help prevent smoke exiting one flue from entering another.

CLEANOUT AND SOOT POCKET

A cleanout is a metal door installed in the bottom of a chimney to allow removal of accumulated soot and creosote. A cleanout is required on all Class A chimneys. A soot pocket is a cavity at the bottom of the chimney to hold soot until it is removed through the cleanout. Each flue should have its own

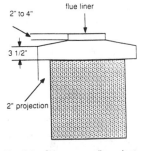

flue liner

2" to 4"

3 1/2"

2" projection

Fig. 7.4 Chimney cap dimensions

cleanout and soot pocket. Cast-iron cleanouts last longer than sheet metal ones.

Cleanouts are manufactured to modular dimensions to minimize cutting. Install a cleanout by leaving the proper void in the chimney wall. Then butter the sides of the void and shove the cleanout into place. Brace it with a tilted block or another secure fixture and do not disturb it until the chimney is finished.

FLASHING

Flashing joins chimney masonry to the building structure to prevent water from entering. Flashing can be constructed on the scaffold or in the shop. Careful cutting and measuring is necessary to produce effective, durable flashing. Each piece of flashing must overlap the ones below it.

Base flashing on the lower side of the chimney is laid on top of the shingles. Base flashing next to and above a chimney must be inserted into the shingles. Counterflashing is inserted into the mortar joints and bent over the base flashing.

In new work, roofers may simplify your work by not laying shingles near the chimney, or only lightly nailing them. Use a flat pry bar if you must remove roofing nails to set the base flashing.

Bend the flashing to fit. To help the mortar bond to the counterflashing, strike the tab that enters the mortar joint with the hooked end of a brick hammer. Then lay the joint with the flashing in place. An alternative is to rake the joint when building the chimney and install the flashing later. Avoid disturbing the flashing while working on the chimney. Flashing should be tuckpointed after the chimney is laid to ensure a complete bond.

Masons may fabricate and install flashings, or they may install flashing fabricated by other trades. Sometimes other tradespeople fabricate and install the flashing (see Fig. 7.5).

1. Cut the first base flashing about 8 inches wider than the outside width of the chimney and long enough to extend about 8 inches up the chimney. This flashing rests on top of shingles below the chimney and is inserted under the shingles beside the chimney.

2. Install subsequent sections of base flashing (rectangles folded at a right angle along their centerlines) as the chimney is built up. Insert them under the shingles and slide them as far uphill as possible.

3. Install the counterflashing for the lower face of the chimney into the bed joints.

4. Cover the lower corners with counterflashing.

5. Continue inserting counterflashing into the chimney as you progress. Insert the final piece of base flashing into the shingles above the chimney and counterflash it.

6. Solder the flashing and counterflashing (if they are galvanized steel) to ensure stability and prevent water penetration. Finally, apply roof cement to any areas that you disturbed.

A "saddle" or "cricket" is a raised portion of flashing uphill from the chimney that prevents water from accumulating at the base. The saddle is used if the chimney's width (parallel to the eave) is more than 30 inches (see Fig. 7.6).

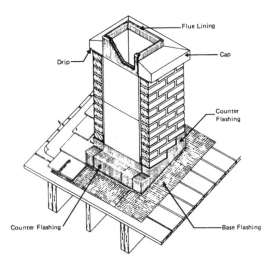

Courtesy of Brick Institute of America

Flue Lining

Cap

Drip

Counter Flashing

Counter Flashing

Base Flashing

Fig. 7.5 Flashing around a chimney

Cricket Dimensions

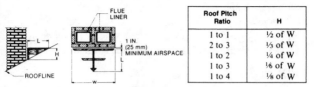

Roof Pitch Ratio	H
1 to 1	½ of W
2 to 3	⅓ of W
1 to 2	¼ of W
1 to 3	⅙ of W
1 to 4	⅛ of W

Fig. 7.6 Cricket framing and dimensions

FLUE

The flue is the channel within a chimney which allows smoke to leave the combustion chamber and building. The flue is generally constructed with flue liner, also called flue tile, a fired clay product that is manufactured in round and rectangular sections that are usually two feet long.

Some codes allow more than one appliance per flue, but in most cases each appliance needs its own flue. A 4-inch nominal masonry wall is required around each flue liner and between adjacent flues. If no liner is used, use an 8-inch nominal masonry wall.

Flue liner should extend at least 2 inches above the chimney cap. If multiple flues are used, they should terminate at different heights to prevent smoke from one flue from entering another. Seal the flue to the cap with flexible sealant.

Sizing flue. Flue size depends on the application. For most residential furnaces, an 8-inch × 8-inch or 8-inch round flue is more than adequate. For commercial furnaces, follow the manufacturer's specifications. Fireplace flue size is largely determined by the opening size. Remember that other factors affect the flue size required:

- *The height of the flue as measured from the smoke chamber to the top of flue. Longer flues draw better than shorter ones and the flue may not need to be as large.*

- *Bends in the flue reduce the draw and may require an increase in size.*

- *Round flues, which allow the smoke to spiral, draw better than rectangular flues with the same inside area.*

- *Conditions outside the building affect the draw. Overhanging trees or buildings can cause downdraft. Steady winds and cold temperatures increase the draw.*

To compensate for these variables, use your judgment and experience when sizing fireplace flues (see *Fireplace*, p. 128).

FOOTINGS

To prevent unequal settling, fireplace footings are generally poured separately from the building foundation. Chimneys are never permitted to rest on wood framing. The footings should extend below frost line and be large enough to support the chimney. Soil beneath the footing should be removed carefully. Do not excavate more than necessary. If voids must be filled below the footing, use sand or gravel. Compact loose soil with a tamper.

HOOD

Chimney hoods are built to prevent downdrafts, rain, and animals from entering the chimney. Hoods are often installed on old construction. The usual procedure is to use a ready-made product. Follow manufacturer's directions.

Masons can build chimney hoods themselves, using either a flat or an arched, ornamental design. No standards exist for sizing hoods, but you should provide at least a 6-inch clearance between the top of the flue and the hood. Flat hoods can be cast in place or on the ground. Arched hoods should be supported with a wooden center during in-place construction.

RACKING

Racking is used to reduce a chimney from the dimension of the fireplace to the dimension required to hold the flue. Make sure to cover the cores of units in the racking. It is best to lay a surface of pavers over the racking. If using a mortar wash, make each step an individual unit for greater durability.

THIMBLE

A thimble is the point where an appliance flue enters a chimney. Thimbles are available ready-made for a round flue, but must generally be sawed into a rectangular flue. Do this by transferring the thimble size to a cardboard template and marking the opening on the liner. Use a portable masonry saw and brick hammer to carefully cut the opening slightly larger than the template.

The joint between the masonry and thimble must be securely seated in fireclay. Eight inches of flue should extend below the thimble to provide a soot pocket.

For wood-burning appliances, the thimble must be located far enough from any ceiling that has no protection from heat. The general practice is to allow at least two diameters of stove pipe beneath the ceiling. The area around the thimble must also be fireproof. This can be done by removing combustible materials (framing, lath, etc.) from the wall and building a support across the wall below the thimble. Then mortar around the thimble for the required distance. Consult building codes.

Be careful to allow full passage for smoke when installing the thimble. The thimble must be flush with the inside of the flue. A thimble that is too deep will interfere with the passage of smoke; a thimble that is too shallow will allow excess heat to enter the chimney wall itself.

LAYING A CHIMNEY

Follow this general procedure to lay up a chimney for a purpose other than a fireplace, while also observing blueprints and building codes:

1. First lay out the walls, then lay out the chimney in the proper location. If the chimney will be built of the same units as the wall, determine the course spacing for the building, and use that spacing for the chimney.

2. Lay out the chimney dry, trying to ensure that only whole units will be needed. Measure the inside area to check that it is sufficient for the flue.

3. Lay the bottom course, bonding all joints with full mortar. Use a 2-foot level to ensure that corners are plumb and that the walls are square and level. To prevent mortar droppings from adhering to the bottom of the soot pocket, lay a piece of bag or cardboard on the bottom and weight it into place.

4. Install the cleanout after building the correct number of base courses. Use bats and whole bricks to fashion the opening.

When the opening is complete, butter mortar around it and force the cleanout into place. Brace the cleanout with a concrete block or other brace. Make sure not to disturb the cleanout during subsequent work.

5. Install the flue liner. Generally, the liner must begin at least 8 inches below the thimble (check codes). The liner should rest solidly on corbelled units as shown in Fig. 7.7.

6. Some masons lay flue tiles with a tight, thin joint of refractory cement before laying the chimney surrounding them. However, it is difficult to place the flues accurately before the bricks are laid, so other masons lay the flues into a section of completed chimney. The inside of the tile must be smooth to prevent buildup of waste products and simplify cleaning. Make sure no mortar from chimney joints touches the flue liner. Leave about 1/16-inch air space to allow for expansion as the flue heats up.

7. Install the thimble flush with the inside of the flue. Thimbles are usually located in the center of the liner and 8" above the bottom of the bottom section. Make sure proper clearances are provided between the thimble and walls, floors and ceilings.

8. Anchor the chimney per blueprints. Anchors are usually provided at each floor level and at the roof line. Generally, a

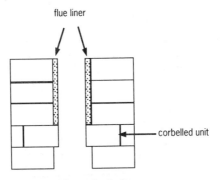

Fig. 7.7 Supporting the flue

1/4-inch steel anchor is embedded into the bed joints and bolted to floor joists or other solid framing (see Fig. 7.3). Some chimneys are anchored with straps around the exterior.

9. Install flashing and insert counterflashing into chimney joints to overlap the flashing.

10. Lay the chimney to full height. Install the cap or cast it into place. Seal flue to cap with flexible sealant.

11. Remove the bag covering the bottom of the soot pocket. Clean the pocket and the entire project.

12. Conduct a smoke test. Close off the top of the chimney and force smoke into the cleanout or thimble. Inspect the chimney to ensure that no smoke leaks from it.

TROUBLESHOOTING

Chimneys are subject to several types of damage:

1. Decay due to extreme weathering. This is most prominent on the section above the roof. Repair by pointing if damage is too extensive. Lack of bond between units indicates the need for a rebuild.

2. Plugging or creosote buildup. Use good firebuilding practices and clean regularly.

3. Chimney fire. Inspect carefully and repair before returning the fireplace to service.

4. Cracked and leaking caps. Seal or replace the cap.

See also *Fireplace, Safe Operation,* p. 140.

CREOSOTE AND CHIMNEY FIRE

Creosote is the common name for chemicals found in woodsmoke that does not completely burn in the firebox. Some creosote typically condenses on the walls of a flue, forming a sticky, black substance that is flammable and hard to remove.

When accumulated creosote burns inside a flue, the condition is called a chimney fire. Chimney fires emit gray or black smoke and can burn a long time at temperatures hot enough to crack the masonry and ignite the structure.

Creosote buildup is worst when:

- *The chimney is exposed to cold outside air.*
- *Long, slow fires are burned.*
- *Green wood is used.*
- *The fire has inadequate burning air because of design or operation.*
- *The chimney is not cleaned often enough.*

Creosote buildup can be corrected by thorough cleaning of the chimney and then removing the cause of the problem. A warm chimney reduces creosote condensation. Follow these steps to reduce condensation and the hazard of chimney fires:

- *Insulating between the backup and the firebox, or using an inside chimney.*
- *Wrapping a chimney running through an attic with noncombustible insulation.*
- *Leaving 2 inches of air space between fireplace, chimney, and wood framing.*
- *Using flue tiles and mortaring them with refractory cement.*
- *Providing an air space around the flue tile and supporting it adequately.*
- *Observing the other cautions listed above.*

Inspect a chimney after a fire to ensure that it is sound. Do this by lowering a light into the chimney and using a mirror. If necessary, remove walls to look for cracks in the chimney.

DOWNDRAFT

Downdraft is a condition in which smoke comes down the chimney in puffs. It is an intermittent problem, unlike smoking, which involves a continual emission of smoke. Downdraft has several possible causes:

- *Inadequate chimney height.*
- *A chimney located under a tree or roof which deflects winds down the flue.*
- *A chimney located downwind of an obstruction that causes an eddy.*

Corrections include installing a chimney cap, increasing the height of the flue, and trimming overhanging trees.

SMOKING

With today's practically airtight houses, draft can be reduced because the exhausting of smoke from the house creates a partial vacuum. The chimney, in effect, is unable to find enough air to dispose of, which results in smoking. The problem is worst during relatively warm, humid days with little wind. Use of ventilating fans in a kitchen or bathroom can worsen smoking.

The solution depends on the problem. First check for flue obstructions by using mirrors and a flashlight. Then test for air availability. If opening a window stops the smoking, the homeowner has two options: opening a window when using the fireplace, or installing an external air intake. Because an air intake requires drilling or sawing through a considerable amount of masonry, homeowners often choose to just open a window.

If opening a window does not help and the flue is clear, the flue is probably too small in proportion to the opening. The opening can be reduced by adding to the height of the hearth or installing a new lintel below the existing lintel. Glass doors will reduce smoking, although not when the fire is being tended. Test any solution before making a permanent alteration.

Fireplaces with multiple openings are most prone to smoking. In extreme cases, it may be necessary to close off one opening. Installation of glass doors on both openings, and opening only one set of doors at any time, may solve the problem.

PART FOUR

THE ABC'S OF MASONRY

8
Skills, Techniques, and Procedures

This section details the techniques—preparing a site, laying out structures, laying units, and cleaning, coating and finishing—that enable masons to construct a wide variety of structures in a safe, efficient, and effective manner.

ANCHORS

Anchors are metal fasteners used to connect masonry to various building materials, such as wood, steel, insulation, and other masonry. Anchors are also used to reattach poorly bonded wythes in cavity walls. Manufacturers have developed many systems to accomplish the difficult task of fastening to masonry. With the variety has come confusion about deciding what device to use for which purpose. For help with this, see *Materials, anchors*, p. 33.

Other items that are sometimes called anchors are described under *Wall ties*, p. 90 and *Joint Reinforcing*, p. 84.

Anchors fall into five general types:

1. Angle bolts.
2. Expanding anchors.
3. Adhesive anchors.
4. Anchors that are driven into place.
5. Reanchoring systems.

ANGLE BOLTS

Angle bolts, used to secure sills and plates to masonry, are installed during construction. Angle bolts for residential construction are generally 3/8 inch diameter by 6 inches to 15 inches long. They have a hooked end to grip the masonry, grout, or concrete in which they are embedded. Building codes specify the size and spacing of angle bolts, but a good rule of thumb is to space them no more than 4 feet apart for anchoring sills and plates. One bolt should be located within 1 foot of each end of plates and sills.

In concrete, angle bolts are installed and braced before the wall, footing, or slab is poured. In masonry, angle bolts are installed in cores or cavities and grouted or mortared in place. It may be necessary to chip the adjacent units to allow a bolt to fit.

Use this procedure to install 15-inch angle bolts into block cores:

1. Lay mesh on top of the cores at the anchor locations in third course from the top. This mesh will support the grout that will be poured into the top two courses.

2. Lay the next-to-last course.

3. Lay the last course, making sure cores that will receive angle bolts line up with cores in the previous course.

4. Grout the cores with grout or mortar as specified.

5. Insert the bolt through the top course to full depth, making sure that grout completely surrounds the bolt and the bolt has the correct depth and is perpendicular to the surface.

6. Cut the grout or mortar flush to the top and protect bolts from disturbance while they set.

EXPANDING ANCHORS

Expanding anchors are inserted in drill holes and expanded into position. You can save work by limiting the hole to the required depth with a stop or another type of indicator when drilling. Make sure drills are sharp; use a hammer drill only with drills rated for use with such a drill. Make sure to match the drill size to the anchor size. When inserting an anchor in a mortar joint, make sure the joint is completely filled, and do not tighten excessively. Otherwise, locate anchors in solid masonry.

Expansion anchors require a solid base material for maximum effectiveness. Wedge-type anchors create a large stress on the base material and are best suited to solid concrete, not unit masonry. Sleeve anchors have a larger bearing area and are more suitable to unit masonry. Drop-in and self-drilling anchors are usually not recommended for use with masonry.

Expansion anchors are not suited to vibratory loads, which tend to loosen them. Use the recommended torque when setting expansion anchors—excessive torque can crack the base material.

ADHESIVE ANCHORS

Cemented or epoxied anchors achieve a chemical and physical bond to sound concrete or masonry. These anchors are suited to heavy-duty applications, such as anchoring machinery, equipment, parking meters, fences, railings, motors, pulleys, and bolts.

Observe the following limitations when using adhesive anchors:

1. Dust and debris must be removed from the hole.

2. While the adhesive can fill small voids, it might not fill large ones. Large voids can reduce performance.

3. High temperatures can reduce the bond.

4. Adhesive anchors are subject to degradation by chemicals. Check chemical conditions before using them.

Use this procedure (see Fig. 8.1) for anchoring a bolt into sound masonry material with hydraulic cement:

1. Drill a hole large enough to accept the bolt and clean out the dust.

2. Place the bolt in the hole and brace it at the exact depth and position.

3. Mix the hydraulic cement and place it in the hole. Follow the manufacturer's instructions for vibrating or slushing the bolt into place.

4. Protect the bolt from disturbance and allow the cement to cure for the proper time before loading it.

A similar procedure can be used for epoxy. Mix the epoxy and pour it in the hole, then allow the anchor to set for the required time. Some epoxies are sold in ready-measured containers which are inserted in the hole and mixed with a hammer drill. See the manufacturer's instructions for details.

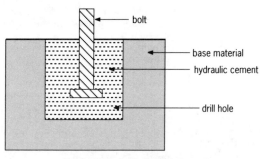

Fig. 8.1 Anchoring in hydraulic cement

DRIVEN ANCHORS

Nails are a crude method of fastening that is suitable for attaching light loads, such as furring strips, to mortar or concrete. Nails are best used with fresh concrete and are not suitable for loads subject to vibration.

Powder-actuated anchors are shot into the base material with a powder charge. Many styles and lengths of nails and studs are available. Minimum penetration is about 3/4 inch in masonry.

Safety: Powder-actuated anchoring tools must be operated by trained personnel. Inspect the tool daily. Do not load the tool until just before use. Wear a face shield and hearing protection. Never point a loaded tool toward any person. Use only with approved charges, studs, and nails.

REANCHORING SYSTEMS

Follow manufacturer's instructions when using a reanchoring system. Certain systems are suited to specific wall types, so choose the device carefully. Make sure to draw up anchors snug but not so tight that they cause further cracking in the wall.

APPRENTICESHIP PROGRAM

Masonry apprentices must pass through a three- or four-year program which includes 4,000 to 6,000 hours of on-the-job training and about 400 hours of classroom instruction. All work is performed under the supervision of journeymen. Apprentices are paid a percentage of the journeymen's pay which increases as the apprentice gains experience.

Topics in this combination of classroom instruction and on-the-job training include materials, tools and equipment, trade arithmetic, plan and blueprint reading, construction details, shop practices, estimating, prefabricated panels, insulation, and safety. A typical program would include the following modules (times are approximate and vary by location and the availability of jobs to provide appropriate training):

150 hours: Safe and proper use of tools, equipment and materials.

2,250 hours: Laying and bonding brickwork, footings, foundations, walls, steps, arches, fireplaces, columns, corners, and welding.

450 hours: Cutting and laying rubble and ashlar stonework.

150 hours: Pointing and cleaning masonry by chemical and mechanical means.

1,275 hours: Setting concrete block, glass block, artificial stone and tile.

225 hours: Waterproofing and fireproofing structures.

Upon successful completion of the program, the apprentice receives a certificate of completion of apprenticeship and becomes a fully trained and qualified journeyman.

ARCHES

Before arches were invented, openings were limited by the strength of lintel stones, and this interfered with building design because many columns were needed to support roofs. The development of masonry arches enabled builders to span long openings.

An arch transfers the weight of the structure above to the piers, or abutments, alongside and beneath it. For an arch to work properly, the 1) span must remain constant, 2) the abutments must sustain the lateral and vertical loads, and 3) the angle of the skewback must remain constant.

Minor arches have an opening less than 6 feet; major arches are larger. Most arches are made with standard-shape bricks, but wedge-shaped bricks are available on special order. These bricks eliminate the need to give the mortar joint a wedge shape.

Arches are described by a specialized terminology:

Abutment: the masonry structure that supports the end of the arch. Sometimes called a pier.

Arch axis: the center line of the curve in the arch.

Creeper: a unit in the wall above the arch that is cut to conform to the extrados.

Crown: the peak of the extrados.

Depth: the distance from intrados to extrados.

Extrados: The curve of the top side of the arch.

Intrados: the curve of the bottom side of the arch.

Radius: the length of a line from a radius point to the intrados.

Radius point: the center of the arch (or a particular part of an arch).

Rise: the height from the spring line to the peak of the intrados.

Skewback: the slanted surface where the arch bears on the abutment.

Soffit: the underside of the arch.

Span: the horizontal length between each abutment.

Spring line: The meeting point between the skewback and the soffit (on a minor arch). The meeting point of the skewback with the arch axis (on a major arch) (see Fig. 8.2).

The most common types of arch are the jack, segmental, semicircular, multicentered or elliptical, tudor, gothic and parabolic. Each may be laid with whole bricks or bats (see Fig. 8.3).

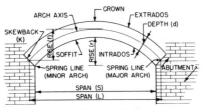

Fig. 8.2 Arch terminology

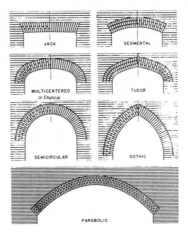

Fig. 8.3 Arch types

Brick Institute of America

LAYOUT

Blueprints or specifications should list the following elements of an arch:

Span and rise

Thickness of mortar joint

Number of courses high

Bonding pattern (number of bricks in one course of the arch)

Skewback angle

Brick size and type

Radius (for a semicircular or segmental arch)

Radius point (for a segmental arch)

Arches are laid over a wooden form called a center that is constructed from plywood cut to the contours of the intrados. Slats can be nailed across the top of the plywood to support the bricks. If a carpenter builds the center, a mason should check its accuracy before installing it.

A basic procedure is described below for beginning the center layout. This procedure is followed by instructions for finishing the center for semicircular, elliptical, segmental, and gothic arches.

LAYING OUT A CENTER—BASIC PROCEDURE

1. Select two pieces of 3/4-inch plywood that are as wide as the span and taller than the rise (see Fig. 8.4). Lay one piece of plywood on a flat surface.

2. Snap line (AB) parallel with the bottom, and about 8 inches above the bottom. (NOTE: This procedure will build centers with extra plywood at the bottom for strength and simplicity of layout. If extra plywood is not wanted, simply place line (AB) at the bottom of the plywood.)

3. Find the middle of (AB) and mark it (C).

4. Mark line (CD) perpendicular to (AB).

5. Measure the rise along (CD) and mark it (E).

For a semicircular arch:

6. Take a string with radius (CE) and draw the intrados, arc (AEB). (In a semicircular arch, the radius equals the rise.)

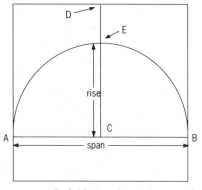

Fig.8.4 Basic arch procedure

For an elliptical arch:

A good approximation of an ellipse can be drawn with this procedure (see Fig. 8.5). Follow basic procedure through step 4. Then begin here:

5. Divide the span by 2. Measure this distance from (C) and mark (E) on (CD). Distance (CE) is the rise.

6. Measure from point (E) 1.25 times the span to line (AB). Mark the intersections (F) and (G).

7. Drive nails into points (E), (F), and (G). Tie a string around these three nails to form a tight loop, (EFG).

8. Remove nail at (E) and put a pencil into the loop you have just formed. Mark the arch by swinging around the intrados with the pencil. Keep the string tight. (FGH) and (FIG) represent alternative locations for the string and pencil during the layout.

For a segmental arch:

Follow basic procedure through step 5 (see Fig. 8.6). Then:

6. Extend line (CD) to point (F).

7. Refer to blueprints or the architect to find either 1) the radius or 2) the distance between the radius point and line (AB). If radius is given, measure one radius from (E) along (EF) to find radius

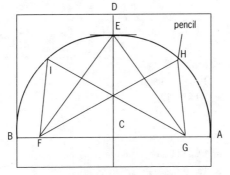

Fig.8.5 Center for elliptical arch

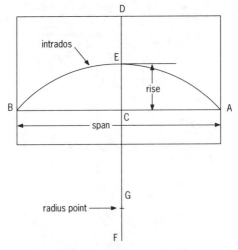

Fig. 8.6 Center for segmental arch

point (G). If length from the radius point to (AB) is given, measure this length and mark radius point (G).

8. Place a nail in the radius point and use a pencil and string to mark the intrados, arc (AEB).

For a Gothic arch:

Do not use the basic procedure.

1. Select two pieces of 3/4-inch plywood at least as wide as the span and as tall as the rise. Lay one piece on a flat surface.

2. Mark line (AB) parallel with the bottom of the plywood. Find the center of (AB) and mark it (C).

3. Mark line (CD) perpendicular to (AB).

4. Measure the rise along (CD) and mark it (E).

5. Measure one-half the span from (C) and mark points (F) and (G) on line (AB). (F) and (G) are the spring points, the ends of the intrados. Distance (FG) is the span.

6. Draw lines (EF) and (EG). Mark (H) at the midpoint of (EF) and (I) at the midpoint of (EG).

7. Draw a line perpendicular to (EF) from point (H) and mark (J) where it intersects (AB). In a similar fashion, draw line (IK) perpendicular to (EG) from point (I). (J) and (K) are the radius points.

8. Place a nail in (J) and in (K). Attach a string of length (FJ) or (KG) (they should be equal to each other and to the radius given by the architect) and tie a pencil to the end. Mark the intrados for each side using this string (see Fig. 8.7).

FINISHING THE CENTER, AND MARKING AND BUILDING AN ARCH

1. Check that the rise, span, radius, and intrados of the completed center are correct.

2. Place one plywood piece on top of the other and tack them together. Cut carefully along the intrados.

3. Nail blocks of lumber between the two pieces. The blocks should be sized so the faces of the plywood will be roughly flush

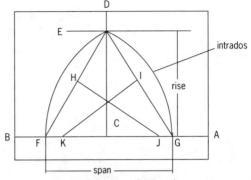

Fig. 8.7 Center for a Gothic arch

with the faces of the finished arch. Lay the center down on a flat surface.

4. Stand bricks on their ends along the intrados. Joints should be relatively thin on the bottom, so the wedge is not too thick at the top. Check the joint spacing to see whether bats or splits will be needed. Refer to specifications or blueprint to find correct number of bricks in the arch. (Arches usually have an odd number of bricks, to allow a keystone at the top, but this is not necessary for structural purposes.) Layout can also be done when the center is in place (see step 6).

5. When the mortar joints are equal and the keystone (if used) is at the top center, mark the joint locations on the center with a pencil.

6. After the abutments are laid up to the spring line, brace the completed center into place with wood posts or dry masonry units. Place wedges beneath the center for easy removal. Make sure the spring points are properly aligned with the abutment. If you made the center with extra material, it will be taller than the rise, so you should brace it with the bottom below the spring line.

7. Check that the center is plumb, level, and correctly located. Use additional bracing to prevent the center from shifting while the mortar sets.

8. If you did not determine joint thickness and the need for bats or splits while the center was on the ground, do so now. Lay out bricks dry and use adjustable shims to space them properly. Then mark the joints on the center with a pencil.

9. Start laying bricks from the sides, making sure joints conform to the pencil lines. Finally, add the keystone (which may be a special stone or brick).

10. Continue laying the wall by cutting creepers (irregular units that match the top of the arch to the wall pattern) on the masonry saw. Each creeper takes its own angle and dimensions.

11. Leave the center in place until the mortar is fully set up and the bricks are self-supporting. This can take just a few hours if the only load is the weight of the wall, although some authorities recommend the center remain in position for as long as a week. After the center is removed, rake 1/2 inch of mortar from the soffit joints and tuckpoint.

BARBECUE AND OUTDOOR FIREPLACE

Barbeques and outdoor fireplaces make excellent gathering places in yards. Their design and construction requires the mason to combine weatherproof techniques with techniques of building fireplaces.

Ready-made fireboxes are available for barbecues, but a grate is the only necessary accessory. If using a ready-made unit, its dimensions will govern the size and proportions of the structure. Provide adequate clearance so metal parts can expand with heat.

FOUNDATION

If possible, select a site that is above standing water all year long. Some grading and added gravel may be needed to ensure good drainage at the site. The foundation should be at least 1 inch wider and longer than the fireplace or barbecue. If desired, provide an apron in front for working at the installation. In areas not subject to frost, a well-tamped surface is enough to support the fireplace slab. In areas with freezing, the foundation must reach below frost depth. You can also float a reinforced concrete slab on the ground. This slab should be at least 6 inches thick, and reinforced with mesh or bars 1 inch from the bottom.

CHIMNEY

No chimney is needed for a fireplace that will only burn charcoal, but wood and coal-burning units need a chimney. An excessively large chimney will distort the proportions of an outdoor fireplace, so make the chimney no larger than needed. The cross-section of the flue is generally no larger than the grate on the fireplace. In some cases, a smaller flue is adequate (see *Chimney*, p. 141).

BLUEPRINTS, SPECIFICATIONS AND SCHEDULES

Blueprints are drawings that give size, shape, and construction details. Specifications are written instructions that describe the quality of work, types, and grades of materials used, which contractor is responsible for various aspects of the job, and how the work will be performed.

If the specifications differ from the blueprints, consult the engineer or architect before proceeding. Such conflicts can arise because:

- *The architect used standard blueprints and customized specifications, without ensuring that they matched (or vice versa).*

- *A change made in an element described in both specifications and blueprints was omitted from one document.*

BLUEPRINT CONVENTIONS

Each drawing is labeled in the title box, usually in the lower right corner. The title box is the first place to look when examining a blueprint.

Several categories of drawings are used to describe various aspects of a project: architectural, structural, mechanical, and electrical. On some residential projects, the architects combine the structural and architectural and even mechanical information on one print. These abbreviations are used to describe the categories:

A: Architectural

S: Structural

M: Mechanical

E: Electrical (used only for larger projects)

Masons always refer to structural and architectural prints, and sometimes to mechanical and electrical prints. Mechanical and electrical prints are helpful for checking details, such as the location and dimensions of chases for pipes or conduits.

The scale of the drawing is commonly expressed in inches per foot. A scale of 1/2 inch per foot means that each foot of building occupies 1/2 inch on the print—and that 1 inch on the print represents 2 feet of the building. Because blueprints are drawn to scale, you can read ("scale") the dimensions of a part from a print with a mechanical drawing ruler. However, if you are unsure of a dimension, do not attempt to scale it from the drawing. Either subtract intermediate dimensions from larger ones to find dimensions omitted from the prints or ask the architect or engineer to clarify the situation.

Drafters use a variety of types of lines to show different elements in a blueprint (see Fig. 8.8):

Object lines show actual structural elements of buildings, such as perimeters of walls and windows.

Center lines divide some element of a building (a fireplace, for example) in two equal parts.

Hidden lines show the edge of an element that would otherwise be invisible in that particular view.

Break lines indicate that the entire element is not fully shown (or was shortened in the middle) to fit the print.

Cutting plane lines indicate where sections are taken (see Fig. 8.9).

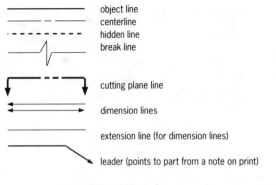

Fig. 8.8 Blueprint lines

Dimension lines show to what points a measurement refers.

Extension lines extend a building element for convenience in showing dimensions.

Leaders are lines which connect notes on the margin to items on the drawing.

VIEWS

Many types of views are used to portray a structure and reduce it from three dimensions to two dimensions:

Orthographic views show structures from a 90° angle. Plans are orthographic views taken from above; elevations are orthographic views taken from the side.

Perspective views show a structure from an angle. They are useful for visualizing a project in three dimensions.

Detail views use a larger scale to show complicated areas, such as footings, fireplaces, chimneys, and steps.

Section views show how a wall is constructed, how various materials are joined, how a wall changes width, or the installation of windows and doors. Section views show how the structure would look if it were sawed along the section line (also called the cutting plane), which is noted on the print (see Fig. 8.9). The section view can identified by a "balloon," usually on an orthographic view. The balloon tells the direction to look at the section, sheet number, and section number.

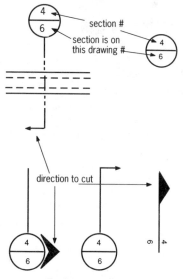

Fig. 8.9 Section lines

PLANS AND ELEVATIONS

Plans are orthographic drawings that show a building from directly overhead. Elevations show the building from a side. Plans and elevations are the builder's basic drawings; the essential ones are:

Plot plans specify where a structure is located on the site. Plot plans are used for laying out the excavation, footings, and foundation, and for landscaping work.

Foundation plans show where and how to build the footings and foundation.

Floor plans describe floors. A separate floor plan is drawn for the foundation, each floor, and the roof.

Elevations show a side of the building and detail the location of windows, doors, and joined structures. Elevations may be named according to which face of the building they represent ("front elevation"). They may also be named by compass directions ("north elevation"). Note that the north elevation faces north, and the viewer faces south when looking at it.

SYMBOLS

Symbols are used to describe masonry, wood, and steel materials in plans. These symbols simplify drawing and reading plans, although there is some variation in the use of symbols (see Fig. 8.10).

Mechanical symbols identify mechanical (plumbing, heating, air conditioning an electrical) elements. These symbols are helpful in

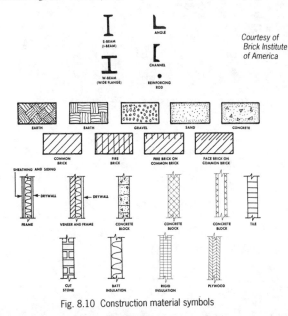

Courtesy of Brick Institute of America

Fig. 8.10 Construction material symbols

locating openings for wires, pipes and other mechanical devices (see Fig. 8.11).

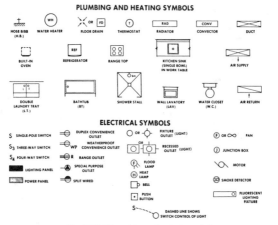

Fig. 8.11 Blueprint mechanical symbols

Brick Institute of America

SPECIFICATIONS

Specifications are written instructions that describe exactly what materials will be used on the job, how they will be stored, in what manner they will be installed, the degree of accuracy required, and the acceptable construction techniques. Specifications accompany blueprints and answer questions that may arise when reading prints.

A complete set of specifications may run to hundreds of pages on a complicated job. Letters on blueprints may refer to elements that are described by specifications.

ABBREVIATIONS

Abbreviations are used on blueprints for convenience. This shorthand varies to some extent from one location to another; see Table 8.1 for a list of common abbreviations.

TABLE 8.1
Blueprint Abbreviations

Term	Abbreviations	Term	Abbreviations
Above Finish Floor	A.F.F.	Divided	DIV
Anchor Bolt	A.B.	Double	DBL
Acoustic	AC	Drawing	DWG
Angle	∠	Drywall	D.W.
Architectural Terra Cotta	ATC	Each	EA
		Elevation	EL
Asphalt	ASPH	End to End	E to E
Basement	BSMT	Existing	EXIST
Beam	BM	Expansion	EXP
Beveled	BEV	Expansion Joint	EXP JT or E.J.
Bituminous	BIT	Exterior	EXT
Block	BL or BLK	Face Brick	F.B.
Bottom	BOT	Facing Tile	F.T.
Brick	BRK	Finished	FIN
Building	BLDG	Finish Floor	FIN. FL
Building Line	B.L.	Firebrick	FB BK
Caulking	CLKG	First Floor Elevation	F.F.E.
Cast Iron	CI		
Cast Stone	CS	Flashing	FL
Ceiling	CLG	Floor	FL
Cement	CEM	Footing	FTG
Center	CTR	Foundation	FDN or FOUND
Center to Center	C/C or C.C.		
Centerline	CL or ₵	Full Size	F.S.
Chimney	CHIM	Galvanized	GALV
Column	COL	Gauge	GA
Concrete	CONC	Girder	GDR
Concrete Block	CONC BLK	Glass Block	GL BL
Concrete Masonry	C.M. or C/M	Glazed Structural Unit	G.S.U.
Concrete Masonry Unit	CMU		
		Grade	GR
Construction	CONST	Grade Line	G.L.
Continuous	CONT	Ground	GRD
Control Joint	C.J. or CONT JT	Head	HD
		Height	HT
Damp Proofing	DP	Horizontal	HORIZ
Detail	DET	Inclusive	INCL
Diagonal	DIAG	Inside Diameter	I.D.
Diameter	DIA or D or ∅	Insulation	INS or INSUL
Dimension	DIM	Interior	INT
Ditto	DO	Jamb	JMB

TABLE 8.1 (cont.)
Blueprint Abbreviations

Term	Abbreviations	Term	Abbreviations
Joint	JT	Rough	RGH
Joist	J or JT	Scale	SC
Length	L or LGTH	Section	SECT
Level	LEV	Sheet Metal	S.M.
Limestone	LS	Siding	SDG
Long	L or LG	Sill	S or SL
Manufacturer	MFR	Similar	SIM
Masonry	MAS	Soffit	SOF
Masonry Opening	M.O.	Specification	SPEC
Material	MTL or MAT	Square	SQ
Maximum	MAX	Standard	STD
Metal	MET	Steel	STL
Millimeter	mm or MIL	Stone	ST
Minimum	MIN	Suspended	SUSP
Miscellaneous	MISC	Symmetrical	SYM
Modular	MOD	Temperature	T or TEMP
Nominal	NOM	Thick	THK
Number	NO. or #	Through	THRU
On Center	O.C.	Tongue and Grooved	T & G
Opening	OPNG	Top of Curb	TC
Opposite	OPP	Top of Foundation	T.O.F.
Outside Diameter	O.D.	Top of Slab	TSL
Over	OVR	Top of Wall	T/W or TW
Overhead	OVHD or O.H.	To Weather	T.W.
Plats	PL or ℄	Tread	T
Position	POS	Typical	TYP
Pound	LB or #	Underground	UG
Precast	PC	Unfinished	UNF or UNFIN
Prefabricated	PREFAB		
Prefinished	PFN	Vertical	VERT
Radius	R	Waterproof	WP
Recessed	R	Waterproofing	WPFG
Reinforced	RNF or REINF	Weatherproof	WP
		Weephole	WH
Reinforced Concrete	R/C	Weight	WT
Reinforced Concrete Masonry	RCM	Welded Wire Fabric	W.W.F.
		Width	W or WD or WDTH
Riser	R		
Rod	RD	With	W/
Rough Opening	R.O.	Without	W/O
Round	RND or ø		

COURTESY OF THE BRICK INSTITUTE OF AMERICA

SCHEDULES

A schedule is a list of materials arranged so you can find information quickly. Doors, windows, lighting fixtures, and materials are commonly listed on schedules. These building elements are keyed to letters on the blueprints; by noticing that door "C" is located in the south wall, you can check the type of door to see if the frame must be installed before laying the wall. Learning to read schedules can save time and mistakes.

TABLE 8.2
Door Schedule

Mark	Size	Description
①	3'0" × 6'8" × 1-3/4"	Steel Entrance
②	3'0" × 6'8" × 1-3/4"	Flush panel exterior
③	3'0" × 6'8" × 1-1/8"	Screen
④	2'4" × 6'8" × 1-3/4"	Flush panel interior
⑤	6'0" × 6'8" × 1-3/4"	Patio sliding glass

BOND BREAK

Bond breaks are building elements that allow adjacent materials to slide against each other. Bond breaks are placed between elements that expand at different rates in response to changes in temperature and moisture. Because metal has a much higher rate of thermal expansion than masonry materials, bond breaks are especially common between metal and masonry. Differential movement as small as 1/4 inch in 15 feet may cause cracking. Bond breaks can be made with building paper or flashing.

BONDS, BRICK

Bonds are patterns of units that are used to create strong and attractive structures. The strength in masonry depends on "breaking the joint"—covering head joints with another unit instead of stacking units. Because common and running bonds cover the head joint most completely, they are the strongest bonds.

Proper bonding requires adherence to these industry standards:

- *The strongest overlap is attained when a stretcher is centered above a head joint in the course below.*

- *Overlaps of less than one-quarter unit are weak and should not be used unless specified by an architect or engineer.*

- *Brick walls without reinforcement should be bonded so at least 4 percent of the surface is headers. Headers should not be further than 24 inches either vertically or horizontally. Metal wall ties or some types of reinforcement may be substituted for masonry bonders.*

- *Headers should be placed at courses where facing and backing wythes are of equal height. Wall ties should follow this rule if possible.*

- *Bats may be needed near corners to establish the proper bonding pattern. Do not use a bat smaller than 2 inches if possible.*

- *Do not allow substances that would break the bond, such as hardened mortar or form oil, to interfere with bonding. However, such "bond-breakers" may be specified at a control joint to allow movement between adjacent sections of wall.*

Plans and elevations of the most common brick bonds are shown in Fig. 8.12.

RUNNING (STRETCHER)

Running bond is the most common bond, and is frequently used in cavity walls, veneers, concrete blocks and tile facing. The bond is composed solely of stretchers. With units that are half as long as they are wide, the head joint is centered on the unit below. With units having different proportions, the head joint may not be centered, which allows the use of complete units at corners. Head joints of alternate courses line up.

Running bond lends itself to use in veneer walls. When used on solid masonry walls, the two wythes are tied together with wall ties or other reinforcement.

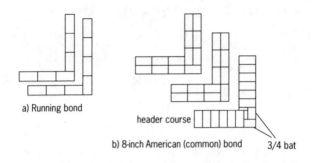

a) Running bond

b) 8-inch American (common) bond

header course

3/4 bat

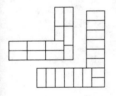

c) English bond

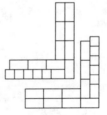

d) 8-inch English bond–bat headers

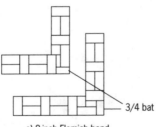

3/4 bat

e) 8-inch Flemish bond

Fig. 8.12 Brick bonds—alternate courses

COMMON (AMERICAN)

Common, or American, bond is used in solid brick walls. The bond consists of five to seven courses of running bond followed by one course of headers, to tie the two wythes together. Each header lines up on a header in the header course below it. A course of Flemish bond, in which headers alternate with stretchers, can be used for the header course. To start the header course correctly, lay two three-quarter bats at the corner, as shown.

ENGLISH

English bond consists of alternating courses of headers and stretchers, with the headers centered on the stretchers. The joints are not broken as effectively as the running and common bonds, but the appearance is very attractive. The header course is started with two three-quarter bats laid parallel with each other at the corner.

English bond can also be laid using bat headers instead of through-the-wall headers. This construction needs wall ties or joint reinforcement for strength. Another variation is "double cross bond." In this bond, a course changes from headers to stretchers, or vice versa, when it turns a corner. When seen from one side only, the bond resembles English bond.

FLEMISH

Each course of Flemish bond consists of alternating stretchers and headers. Flemish bond must be used with solid walls. Corners can be laid with half-bats and three-quarter bats; 2-inch closures; quoins; or other combinations.

STACK

Stack bond is laid with units stacked one atop another, with no attempt to break the bond. Stack bond is weak and mainly used for its aesthetic properties. It is often used for interior concrete block walls and is very common in glass block walls. The bond has good loadbearing qualities and can be strengthened by reinforcing the cores and joints or by adding bond beams. Stack bonding is often used when concrete block is laid with surface bonding cement instead of mortar.

Unit placement is especially important for stack bond because the head joints are so prominent. Use a vertical mason's line every few head joints to maintain proper spacing.

BONDS, PATTERN

Pattern bonds are methods of laying units to take advantage of the esthetic effects of various combinations of color, texture, and placement. To make a pattern bond, units can be recessed or protrude from the surface; unit and mortar colors can be chosen to harmonize or contrast; and units mixed. These esthetic devices can be coordinated with the bond and the tooling for maximum effect.

Pattern bonds are used for both paving and walls. A nominal size of 4 inches × 8 inches is needed for herringbone, square, and some basketweave patterns. These patterns can be laid dry with units having actual size of 4 inches × 8 inches or mortared with units having nominal size of 4 inches × 8 inches. Experiment if you are unsure what size of unit will work in a certain pattern. Stack bond can be laid with any rectangular units, either mortared or dry. Special units, including hexagons and various keyed shapes, are also used for paving and pattern bonds (see Fig. 8.13).

SQUARE PATTERN

Square pattern requires units whose nominal length is twice the nominal width. A square pattern is easy to lay out in square areas or areas that can be divided into squares of various sizes. If the area cannot be divided into squares, start the pattern in one area and fill in borders with fragments of square pattern. In any case, treat each square as a separate panel.

Make the paving conform to the desired slope by laying the corner units for each square at proper height according to the pavement slope. Then fill in the sections using the corners as height guides. Use this procedure to pave with the square pattern:

1. Mark off the sections according to plans or break up an area into sections of nearly equal size. To break up an area, first measure

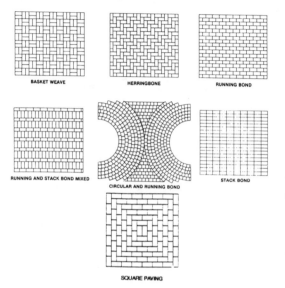

Fig. 8.13 Brick paving patterns

it. Then choose the number of units desired as the side of each square—perhaps 10 or 16—and multiply this number by the nominal length of a unit. For example, 16 units of 8-inch nominal length would produce a square 128 inches on each side. Now divide the total length and width of the paved area by this dimension to find the number of squares required. Repeat with a different number if necessary to find the size of square that will pave the area most efficiently.

2. Start laying out the area by reference to a starting line, perhaps a wall or sidewalk. Mark off squares with crisscross chalk lines. Check that lines are square, parallel, and have the proper spacing. Then spray paint the lines or use another means to be sure they remain visible until the paving is completed.

3. Calculate the slope in terms of inches per foot. Then translate this into slope per square. (If the slope is 1/8 inch per foot, and the square is 6 feet long, each square has a slope of 6/8 inch or 3/4 inch.)

4. Pave the highest square first. Start with the corner units, then lay the outside courses. Use the corner units as guides for height and slope, or attach a bevel to a level to guide you. Work toward the center of the square and do not kneel on freshly laid units. If using mortared paving, fill the joints thoroughly before the mortar sets.

5. Finish all squares and adjoining border sections. Tool joints and clean mortar from the surface. If the paving is dry, sweep sand into the joints.

HERRINGBONE

Herringbone pattern can be either diagonal or square. In square herringbone, units are laid with the bed parallel or perpendicular to the sides of the rectangle they comprise. Diagonal herringbone should be laid in a wall with units set at 45° to horizontal. In a floor, units are set at 45° to the borders of the rectangles they comprise. Use the following procedure to lay out a mortared, diagonal herringbone panel on a floor or wall (see Fig. 8.14):

1. In a wall, establish a horizontal line called the base line for the panel. The base line should be about 6 inches from the bottom of the panel.

2. On a pavement, mark out the square panels according to the plot plan, or use the technique described above for square paving. Locate the base line 6 inches up from the bottom of the rectangle and parallel to it.

3. Lay out the first course dry (see step 1). The top corners of this course should just touch the base line. Use a 45° triangle to get the proper angle. The units will overlap the panel perimeter on three sides. Adjust the layout so small pieces of brick do not overlap the perimeter, to prevent leaving weak chips.

4. Lay the second course dry on the pavement (see step 2). Allow enough room for the joints between courses.

5. Mark the outside lines of the panel on top of the two dry courses.

6. Cut enough units for two courses.

7. Check your accuracy by replacing units in the layout on the pavement.

8. If the cuts were correct, calculate how many cut bricks will be needed for the entire panel. (Cuts should be the same for alternate courses, except for the final course.) Mark and cut enough units for the whole panel.

9. Lay the first course in mortar, using the 45° triangle to align the units accurately. Place the level as shown across the top corners of each unit in the course to control joint thickness. Use a mason's line or level to keep the course flat. Observe proper slope in a pavement.

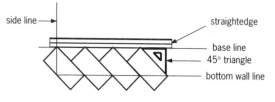

step 1

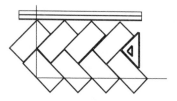

step 2

Fig. 8.14 Diagonal herringbone layout

10. Repeat this procedure for the following courses.

11. Mark top and side cuts for the final course and lay the units in proper locations. Then tool joints and clean the panel as required.

BASKETWEAVE

Basketweave pattern uses groups of either two or three units laid parallel to each other to form small squares. (The pattern shown uses groups of three units.) Each square adjoins other squares whose units are laid perpendicular. If the squares are parallel to the sides of a rectangular panel, the pattern is parallel basketweave. If the squares run diagonally, the pattern is diagonal basketweave.

Use this procedure to lay a square basketweave pattern:

1. Mark off the surface into squares using the procedure outlined for square paving.

2. Establish the slope for the paving area and mark it on temporary stakes.

3. Dry lay two sides to establish the pattern. Then locate groups of three units at each corner in the proper alignment. Use the level and refer to the stakes so the paving remains in proper alignment.

4. Lay one outside line of three units depth, using the corner units as guides. Then work one full line at a time, making sure to maintain uniform joint thickness. Or lay the entire perimeter, three units thick, to provide a guide for the remainder of the work.

5. Finish the pattern, tool the joints, or brush sand into them.

LAYING A DIAGONAL BASKETWEAVE

Use this procedure (see Fig. 8.15) to make a diagonal basketweave pattern bond:

1. Make a template of cardboard. Lay a 45° triangle into the corner of a square of cardboard. Slide the triangle until its hypotenuse is the same length as the nominal length of the unit. Then mark

the hypotenuse and cut the template. The finished template has two 45° angles and a hypotenuse the length of one nominal unit.

2. Mark the squares following the procedure for square paving.

3. Insert the 90° angle of the template into a corner of one square. Mark the hypotenuse on the pavement base or backing wall. This is the "diagonal base line."

4. Mark the "perpendicular base line" at 90° to the diagonal base line.

5. Place a unit with its long side against the diagonal base line. Then lay two more units to make one small square of basketweave pattern.

6. Continue laying groups of three units along the perpendicular base line, alternating the orientation to make the basketweave pattern. Use the level to maintain the proper slope.

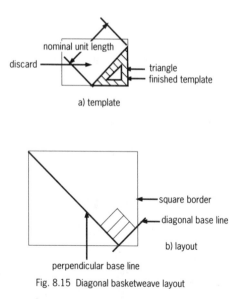

nominal unit length

discard

triangle
finished template

a) template

square border

diagonal base line

b) layout

perpendicular base line

Fig. 8.15 Diagonal basketweave layout

7. Mark other perpendicular base lines if needed to lay all whole units for the pattern.

8. Mark, cut, and lay cut units as needed to fill in around the edges.

9. Tool the joints and clean surfaces.

CAULKING

Caulking is a general term for gunnable sealants used to make a weatherproof joint between two masonry surfaces, or between masonry and other building elements. Caulking prevents water leaks and air drafts, saves energy, and improves appearance. Caulking is also used to seal prefabricated masonry panels and movement joints.

Factors to consider when choosing a sealant include materials to be bonded, cost, durability, color, application temperature, and chemicals in the environment (for industrial applications). The color is generally selected to harmonize with the masonry or the adjacent material.

Specialty sealants are available for these applications:

- *High-traffic areas.*
- *Water seepage areas.*
- *Locations exposed to chemical attack or extreme temperatures.*
- *Movement joints.*

APPLICATION

Before caulking, be sure that surfaces are clean and free of oil and other contaminants. Previous sealants must be removed by wire brushing or scraping. Some caulks require priming first.

Wide gaps must be filled with fiberglass, oakum, foam, or other filler to back up the caulking. In many cases, a certain ratio of depth to width must be maintained. Consult manufacturer's instructions about this and all other details of using a particular compound.

Some materials must be mixed before use, while others must be warmed to proper temperature. Cut the tip of the cartridge at an angle to produce the desired width of caulk, or to reach the bottom of the void, as specified by the manufacturer. Keep the tip clean during work to prevent irregularities in the seal. In vertical joints, work from the bottom up to prevent air spaces in the joint.

After caulking, tool the joint for appearance and better adhesion to the base material. Tooling may be performed with a dry tool or one wetted with a solvent or water. Some caulks may be sealed with a glossy agent, while others may be painted.

CHASES

Chases are recesses provided for mechanical building components, such as pipes, ducts, and wiring. Chases are shown on blueprints. Some may be described on detail views as well.

VERTICAL BRICK CHASE

Vertical chases extend part or all the height of a wall. They are used in walls that are at least 12 inches thick (chases in thinner walls would unacceptably reduce strength). Use the following procedure to lay a vertical brick chase:

1. Read plans and specifications to determine the location, size, and extent of the chase. Make sure the chase is large enough to handle the mechanical equipment it must hold.

2. Mark building corners and lay the first course dry. Measure from the corner to the chase. Be sure to note whether the plans measure to the center line or the edge of the chase. Mark the chase on the foundation.

3. Lay out the chase bond to determine the most efficient arrangement for it. Adjust the bond if possible to reduce cutting. Then mark cuts on the dry laid units and cut them.

4. Lay the wall and the chase at the same time. Plumb the insides of the chase as you work. Make sure not to build the chase any higher than the plans specify.

HORIZONTAL BRICK CHASE

The procedure is similar for a horizontal chase:

1. Read plan, elevation, section, and detail views to determine the location, size, and extent of the chase. Determine elevation of the chase and make sure you are measuring from the proper benchmark.

2. Mark off building corners and the first course. Measure from the corners to mark the horizontal extent of the chase. Mark the chase clearly on the foundation. Mark the height of the chase on the story pole.

3. Lay out the bond to determine the most efficient arrangement. Plan the wall so a header course is on top of the chase.

4. Lay the wall until the course below the chase. Check by measuring the exact location of the chase from the corners.

5. Place extra lines if needed to control alignment of remaining wythes while you lay the chase courses.

6. Place a temporary brace along the full length of the chase to hold the headers until the mortar sets. 2 × 4 blocks supporting a 2 × 4 header make a good brace. Place wedges under the blocks for easy removal.

7. Lay the upper courses. When the wall is fully set, carefully remove the bracing.

BLOCK WALL CHASES

Vertical chases in an 8-inch stack bond course are constructed by replacing one stack of 8-inch blocks with a stack of 4-inch blocks. Check building codes to ensure a chase is permitted in the type of wall you are working on.

CLEANING

Because of the great variety of masonry materials and the number of possible sources of stains, cleaning is one of the more confusing parts of masonry. Cleaning is performed after new work is laid and when renovating old buildings.

The two general categories of cleaning methods are wet and dry. Wet cleaning techniques include acid, cleaning compounds, steamcleaning, and water. In general, wet methods are less destructive to masonry than dry methods. Dry techniques include burlap, grinding, rubbing stone, sandblasting, and scraping. Sandblasting and grinding can damage masonry, so they should be used with caution, generally after other techniques have proven ineffective.

In renovation, dry cleaning is often done before the joints are repaired, and wet cleaning is done afterward. This will prevent cleaning water from penetrating the wall and causing efflorescence or freezing damage. On most jobs, the architect or engineer specifies the cleaning method. Otherwise, consult the masonry manufacturer or a distributor of cleaning solutions or equipment before cleaning.

Because improper cleaning can do more harm than good, you must understand the cleaning agent, the building materials and the problem you are trying to correct. Some mistakes are permanent, so you should always search for the least drastic means of cleaning. Always test a small area before proceeding with a large cleaning operation. Sometimes, different sections of a building need different cleaning techniques. In general, smoother materials are easier to clean than rougher ones.

Cleaning failures have three general causes:

1. The wall was not saturated with water before and after cleaning. This causes the wall to absorb cleaning solution and can lead to mortar smears, efflorescence, or other staining.

2. Improper mixing or use of solution. Overly concentrated acid can damage mortar joints or discolor units.

3. Lack of protection for adjacent materials, such as windows, trim, and shrubbery. Acid solutions will corrode metal trim.

KEEP THE WORK CLEAN

The first step in cleaning masonry is to avoid soiling it. Follow these practices to keep a new structure as clean as possible:

- *Store new units off the ground and underneath plastic.*

- *Cover work at the end of the day.*

- *Cover the bottom 4 feet of the wall as soon as possible with straw, plastic, or plywood. Try to keep this section covered until landscaping is completed. This will prevent falling mortar, grout, dirt, and dust from soiling the wall.*

- *Cut extruded mortar in a manner that prevents smearing.*

- *Wait until mortar spatters on the wall have set, then cut them off with the trowel. Use a stiff brush or burlap to remove remaining mortar scraps. Remove set mortar with a board or broken brick.*

- *Pour grout carefully and from the inside to minimize spills.*

- *Keep scaffold uprights away from the wall so mortar cannot accumulate between them and the wall. Turn scaffold boards on edge at day's end to prevent rain from splashing dust and mortar against the wall.*

WET CLEANING

ACID CLEANING

The most effective solution for removing cement-based products is hydrochloric acid, known to masons as muriatic acid. Muriatic acid can dissolve mortar, calcimine, and whitewash to 1) improve appearance or 2) provide a clean surface for new masonry materials to bond. Muriatic acid is dangerous (see *Safety,* p. 326) and damages joints so much that some masons say a building ages twenty years from a single washing with acid. Clean construction practices and prudent application of acid will reduce this damage, but not eliminate it.

Oxalic acid is used to remove rust stains from structures. Scrub with a hot, concentrated solution. Other cleaning compounds are available for certain stains and building materials. Check with your supplier for specific needs. Some cleaning solutions contain both acid and detergent to remove mortar and dirt. Observe the precautions listed below when using any acid cleaning solution.

The best practice for washing is to test the proposed treatment on scrap bricks or inconspicuous areas before general application. Wait

TABLE 8.3
Cleaning Methods for New Masonry

Brick Category	Cleaning Method	Remarks
Red and Red Flashed	Bucket and Brush Hand Cleaning High Pressure Water Sandblasting	Hydrochloric acid solutions, proprietary compounds, and emulsifying agents may be used. **Smooth Texture:** Mortar stains and smears are generally easier to remove; less surface area exposed; easier to presoak and rinse; unbroken surface, thus more likely to display poor rinsing, acid staining, poor removal of mortar smears. **Rough Texture:** Mortar and dirt tend to penetrate deep into textures; additional area for water and acid absorption; essential to use pressurized water during rinsing.
Red, Heavy Sand Finish	Bucket and Brush Hand Cleaning High Pressure Water	Clean with plain water and scrub brush, or *lightly* applied high pressure and plain water. Excessive mortar stains may require use of cleaning solutions. *Sandblasting is not recommended.*
Light Colored Units, White, Tan, Buff, Gray, Specks, Pink, Brown and Black	Bucket and Brush Hand Cleaning High Pressure Water Sandblasting	*Do not use muriatic acid!* Clean with plain water, detergents, emulsifying agents, or suitable proprietary compounds. Manganese colored brick units tend to react to muriatic acid solutions and stain. Light colored bricks are more susceptible to "acid burn" and stains, compared to darker units.
Same as Light Colored Units, etc., plus Sand Finish	Bucket and Brush Hand Cleaning High Pressure Water	Lightly apply either method. (See notes for light colored units, above.) *Sandblasting is not recommended.*

TABLE 8.3 (cont.)
Cleaning Methods for New Masonry

Brick Category	Cleaning Method	Remarks
Glazed Brick	Bucket and Brush Hand Cleaning	Wipe glazed surface with soft cloth within a few minutes of laying units. Use soft sponge or brush plus ample water supply for final washing. Use detergents where necessary and acid solutions only for *very difficult mortar stain.* Do not use acid on salt glazed or metallic glazed brick. Do not use abrasive powders.
Colored Mortars	Method is generally controlled by the brick unit	Many manufacturers of colored mortars do not recommend chemical cleaning solutions. Most acids tend to bleach colored mortars. Mild detergent solutions are generally recommended.

COURTESY OF THE BRICK INSTITUTE OF AMERICA

as long as possible to check the effects before proceeding. Do not use muriatic acid on light colored bricks, including pink, buff, gray, or bricks with specks of color. Do not use muriatic acid on concrete blocks because it will release metallic elements which will oxidize and discolor. Colored mortar often bleaches under acid treatment.

Although the usual practice is to start washing walls at the top, starting at the bottom will help prevent streaking and reduce acid retention in the wall. Washing from the top down allows spent acid and minerals to drip onto dry wall, where they can be absorbed, and possibly cause efflorescence. If you wash from the bottom up, the spent acid will drip onto wet sections of wall, which are less able to absorb it.

Acid washing procedure

1. Allow complete hydration to take place before washing, but do not wait too long because old mortar is hard to remove (mortar smears and stains over six months old may be extremely resistant to removal).

2. Cover all trim, shrubbery, limestone, and cast stone with plastic.

3. Use a solution of one part acid to ten to twenty parts of water, always adding acid to water and never water to acid.

4. Test the cleaning solution on a hidden area or scrap material. If it does not dissolve the extra mortar, gradually increase the acid concentration to one to ten parts of water. For best results, let the sample age for one week before evaluating it.

5. Soak the wall to prevent efflorescence that can be caused by water penetrating the wall. Use in temperatures above 50°F for best results.

6. Use fiber or plastic brushes to scrub the wall while the acid is foaming. Work on small areas at a time. Start from the bottom and work upward (see above). Allow acid to remain five to ten minutes and rinse. Follow the manufacturer's instructions on proprietary compounds. Rinse the wall immediately afterwards.

Acid safety. Observe proper safety precautions when using acid. Dilute the acid to the proper strength by adding acid to water—not water to acid. Use a face shield and long rubber gloves. Use rubber or plastic buckets and a long-handled brush to keep you away from the work. Clean the wall thoroughly after work. Avoid breathing fumes. Use adequate ventilation if working indoors.

DETERGENTS AND OTHER CHEMICALS

Detergents can remove light films of mortar from units that should not be washed with acid. Many cleaning products are available for specific tasks: lime solvents, restorers for old masonry, smoke, soot and iron removers, and specific compounds for removing various chemical stains or graffiti. Most cleaning compounds should be used with ventilation. Wear goggles or face shield and rubber gloves.

Follow this procedure unless the manufacturer specifies otherwise:

1. Remove dirt and mortar with a stiff brush, trowel, or burlap bag.

2. Soak the wall with water to remove loose particles.

3. Mix the proprietary detergent as specified (some are used concentrated, others diluted) or use 1/2-cup TSP and 1/2-cup household detergent in one gallon of water. Scrub the areas that need cleaning. (NOTE: some municipalities may prohibit the use of TSP.)

4. Flush the wall to clear all chemicals and residue.

STEAMCLEANING

Steamcleaning uses steam sprayed at high pressure (140 to 150 pounds per square inch) against masonry. Unlike sandblasting, this technique does not degrade the surface. Any of several chemicals, including sodium carbonate, sodium bicarbonate, and TSP, may be added to increase the cleaning action.

A small area is steamed and then rinsed with a separate nozzle. Resistant deposits can be removed with a scraper or wire brush and then steamcleaned. Operators should wear protective gear and check equipment carefully before using.

WATER

Water washing should not be performed until at least seven days after the mortar has set. Water washing is the mildest cleaning technique and is preferred if it will work.

Some high-pressure cleaners can also apply solutions of muriatic acid. Observe all safety precautions when spraying acid. Some pressure washers can heat the cleaning solution for increased efficacy. Make sure to provide drainage on big washing jobs or the water might damage the foundation.

DRY CLEANING

BURLAP OR BRUSH

Burlap or a stiff brush can serve as a coarse cleaning tool for nearly hardened mortar. Using either tool at the proper time can save considerable cleaning later on.

GRINDING

Hand-held grinders are used to remove dirt, but they tend to take off the top layer of masonry as well. Grinding also produces fine dust, which must be removed later. Operator protection is vital: wear goggles, ear protection, and mask or respirator.

Grinding should be done cautiously on soft material, where it can leave a pattern of swirls or scores.

RUBBING STONE

One of the simplest techniques for removing scrap mortar is to rub it off with a stone. This can be a special stone or just a broken piece of brick or block.

SANDBLASTING

A sandblaster forces compressed air and sand through a nozzle to blast dirt and foreign matter from a surface. Sandblasting is probably the most effective means of cleaning heavy soil from structures, but it is also one of the harshest, and it can harm both units and mortar. Test the procedure on spare bricks or an inconspicuous area first. The procedure should be done by trained personnel inside covered scaffolds. Some communities have regulations against sandblasting.

Because sandblasting removes some of the surface along with the stains, it can leave a rough finish that will gather soot and soil. After sandblasting, tuckpointing is frequently required to repair joints. Because the surface loses most of its weather resistance, a coat of sealant is also suggested. The sealant will also help repel soot and dust.

Sandblasting is not recommended for bricks with a heavy red sand finish. If possible, use an aggregate other than silica, because silica dust causes serious respiratory diseases.

SCRAPING

Scraping with a trowel or paint scraper can remove mortar after it begins to harden. Wait until the mortar begins to set to prevent smearing, but scrape before the mortar is fully set.

CLEANING EFFLORESCENCE AND STAINS

Use the following suggestions for cleaning efflorescence and stains:

Efflorescence. Remove white efflorescence with dry or wet brushing. For heavy accumulations, mix one part muriatic acid with twelve parts water and follow above directions for acid washing.

Green stain. Green stain is caused by vanadium salts found in some bricks. Green stain can be minimized by storing brick under cover and off the ground, not cleaning light-colored brick with acid, and using recommended cleaning procedures. To remove a green stain, neutralize the acid by flooding the wall with water and spraying a mix of sodium hydroxide (1/2 pound per quart of water). Leave the solution on the wall for a few days and flush it off (see *Troubleshooting, efflorescence*, p. 295).

Paint. Fresh paint can be removed with paint remover. Old paint must be abraded with steel wool, sandblasting, or a similar procedure. Some proprietary compounds are effective with old paint, but test them first.

Smoke. Smoke may be removed with a scouring powder containing bleach or an alkali detergent with an emulsifying agent. Smoke can be difficult to remove.

Dirt. Dirt can be hard to remove from textured surfaces. Try scouring powder and a stiff brush. High-pressure steam cleaning is an alternative for large areas or resistant dirt.

Plant growth. Areas which are constantly damp may show signs of moss, which can be killed by a commercial weed killer or an ammonium sulfate solution.

COATINGS AND MOISTURE RESISTANCE

Moisture is one of the major problems in masonry. Masonry materials lose a great deal of resistance to heat flow when they are wet, thus increasing heating and cooling costs. Moisture can also cause efflorescence, spalling, cracking, and other forms of degradation.

Four factors influence the degree of moisture resistance in a masonry structure: design, materials, construction practices, and maintenance. Each must be correct for a structure to perform its best.

Design must take into account common sources of moisture: soil, rain, and humidity from within the building. The design must specify proper flashings and other details to prevent water entry.

Materials, including units, mortar, and other components, must be chosen to work properly in their environment. The materials must also work together to prevent corrosion and other forms of decomposition.

Construction practices, including proper filling of joints and storage of materials, will help ensure that a design reaches its potential.

Maintenance, including regular inspections, repairs, and coatings, will help ensure that a sound building remains sound.

COATINGS AND FINISHES

Coatings and finishes are applied to new and existing masonry structures to 1) change the texture and/or color, or 2) improve weather resistance. These finishes are effective only if they are applied over a well-prepared surface in accordance with manufacturers' instructions (see *Troubleshooting, water*, p. 294).

Because such a variety of materials is available to increase the resistance of a masonry structure to water, it is impossible to explain how to use each one. Instead, some common procedures are described here.

To prevent walls from becoming wet or saturated, it is a poor practice to apply waterproof sealers to the cold side of a wall, as this will prevent the escape of water vapor. Some exterior paints are breathable and allow water vapor to escape; these paints should not cause unwanted moisture buildup.

Two general categories of materials are used to coat masonry: materials containing portland cement, including cementitious waterproofings, parging, plaster and hydraulic cement; and other materials, including membranes, paint, and clear coatings. These materials are described below.

PREPARATION

The first step when considering whether to coat an existing masonry wall is to inspect it to determine the source of the moisture. During this inspection, examine:

- *Flashing*
- *Weepholes*
- *Copings and caps*
- *Caulked joints*
- *Mortar joints*
- *Condition of details and roof*

Before proceeding with a coating job, make sure to correct the deficiencies you have found. Coatings will not correct the problems listed above. After correcting the problem, wait a few months before coating; there may no longer be a need for the coating.

Once the need for coating is established, you must prepare the surface. New masonry must age for the period specified by the coating manufacturer. Old surfaces must be clean to allow proper penetration of the coating. With some exceptions, the wall must be dry at the time of coating. Temperatures must be within certain limits during the work.

APPLYING CEMENTITIOUS WATERPROOFINGS

Many waterproofing products on the market combine portland cement and aggregate. These long-lasting, strong-bonding materials can be trowelled or brushed on a surface. Similar waterproofing products may be used to form a cove at the junction between a basement floor and foundation, or between the footing and foundation.

This is a general description for using these materials. Consult manufacturer's directions when using specific products:

1. Prepare the surface. Old structures must be clean. Chip, sandblast, or grind defective material. Remove coatings, efflorescence, fungus, and other foreign matter. Repair all cracks and voids. Mortar must be fully cured before coating. Use a bonding agent when mixing coatings to be applied to extremely dense concrete or masonry. Observe temperature limitations on the package.

2. Mix the coating to a thick batter. A brush should stand up in the material when it is ready to apply.

3. Dampen the surface ahead of the application. If the material pulls off the wall, dampen the wall again. Use little or no water in damp, cool conditions.

4. Apply the coating with a brush or trowel. A special brush may be recommended; otherwise use a 6-inch tampico brush, not a paintbrush. A long handle will speed the application and reduce fatigue.

5. Brush or trowel a heavy coat per instructions. Greater water pressure behind the wall indicates the use of a heavier coating.

6. If two coats are needed, apply the second coat before the first has fully set—after about 24 hours in average conditions. For some applications, a different material, such as an acrylic paint, can be used for the second coat.

7. Allow the coating to set. In hot, dry conditions, fog the surface every few hours during the setting period.

PARGING

Parging is the application of a thin cementitious coat to a surface to increase water resistance. The parging material may be the same mortar used to lay the wall or a mix of portland cement and sand. A water-reducing agent can be mixed in to reduce shrinkage of the applied parging.

Typical parging applications include 1) between wythes on a composite wall, called backplastering or backparging, and 2) on the exterior of a foundation.

The joints on a wall to be parged should be cut flush with the trowel during construction. Then apply mortar about 3/8 inch to 1/2 inch deep to the surface with the back of the brick trowel. There is no need to make the parging very smooth, and the application can be done rapidly.

Two coats are preferable on outdoor parging. Work from the bottom up and make the upper coat overlap the lower to improve water runoff. Make sure to fill all cracks and voids completely to prevent water infiltration. Keep the wall damp with a fog spray to improve hydration. A second coat is usually applied within 12 to 24 hours. Cover the parging and keep it damp for about 48 hours, depending on conditions.

PLASTER AND STUCCO

Plaster and stucco are both compounded of portland cement and sand. They are essentially the same material, although stucco usually refers to an exterior coating while plaster may also be used inside. The terms are used interchangeably here. The advantages of portland cement plaster stem from its hardness, fire resistance, durability, and low cost. Many finishes and colors are possible.

MOVEMENT JOINTS

Cracks in stucco can arise from 1) movement in adjacent structures, such as ceilings or plumbing fixtures; 2) settling of a foundation; and 3) cracks or joints in underlying structures.

Prevent cracks by:

- *installing movement joints in the plaster over movement joints in the substrate.*

- *installing movement joints every 20 feet in either direction in a wall or ceiling.*

- *ensuring adequate clearance around pipes, ducts, and other building elements that may shift independently from the structure.*

SURFACE PREPARATION

Plaster must be workable, capable of adhering to the substrate, and not likely to sag while curing. Cement can be portland cement, Type M masonry cement, or a mix of lime, portland cement, and sand. Mixing and workability requirements for plaster are similar to those for block and brick mortar. Do not add excessive water, which can cause crazing, low strength, efflorescence, and other problems.

Plaster must bond securely to the substrate, with chemical and/or physical bonds. New concrete block makes an excellent substrate because it allows both types of bond. Test old concrete by spraying water on it. If the water absorbs readily, the plaster will probably bond to the surface. If not, apply a "dash-bond" coat. This material is one part portland cement and one or two parts of sand, with enough water to create a paint-like mix. Various bonding agents can also be used; follow instructions from the manufacturer.

Make sure the surface is free of oil, dirt, or foreign matter that will hamper the bond. Painted masonry can be sandblasted to create a good substrate. A layer of metal lath over roofing paper will provide adequate bonding over any sound surface. Tack up the roofing paper and anchor or nail the metal lath to the framing.

APPLICATION

Spray water over an absorbent backing before applying plaster. Two or three coats of plaster are applied, depending on the application and method of covering. Three coats are required over a metal lath base. Two coats are permissible over a sound masonry base and on horizontal surfaces.

Plaster is either trowelled with a plaster trowel or machine-applied. Machine application is faster and more uniform in texture and color, and leaves no joint or lap marks. With machine application, add color to the two final coats. One coat ensures total color coverage and the other creates the desired texture.

CURING

Moist curing is needed to develop a strong plaster coat. In cold conditions, use practices recommended for cold weather masonry. ANSI Specification A42.2 calls for curing in a enclosure kept at 40°F for 48 hours before and after application.

In normal weather, cover the work with plastic for 48 hours. In hot weather, use tarpaulins or plywood barriers to reflect sunlight and keep the plaster cool. Fine mists of water can be substituted for the plastic covering during hot weather. Fog spray is best if continuously applied during the curing period.

FILLING A LEAKING CRACK

Hydraulic cement, which sets quickly under water, can be used to repair cracks in basements, tunnels, and shafts. For cold conditions, warm the repair area and mixing water to hasten the set. For warm conditions, use cold water to retard the set.

Compare the following procedure to manufacturer's specific directions before use:

1. Prepare the surface. Remove defective mortar and concrete from the crack. Cut cracks at least 3/4 inch wide and deep. Remove any tie wires or other metal embedded in concrete to a depth of 3/4 inch.

2. Mix the hydraulic cement with water. Warm water will accelerate the set; cold water will retard it.

3. Force the cement into the crack or joint. If water is streaming out, hold the cement until it becomes warm in your hand, indicating that it is starting to set. Then force it into the crack and hold it with a trowel until it hardens (usually within 3 to 5 minutes).

4. Cut off the repair flush with surrounding surfaces.

5. Scratch the surface before its final set if you will apply a finish coat over the patch.

PAINT
PORTLAND CEMENT

Portland cement-based paints have excellent durability when applied to masonry. These paints form a hard surface but are permeable to water vapor, so moisture buildup should not be a problem. Portland cement paint is available prepared or it can be mixed on the job. Prepared paint offers greater consistency of results.

Two types of portland cement paint are available:

Type I contains at least 65 percent portland cement by weight. It also contains 20 to 40 percent siliceous sand for use on surfaces with open textures.

Type II contains at least 80 percent portland cement by weight and is more durable than Type I.

Portland paint can be applied to fresh concrete, but three weeks of curing is preferable. Moisten the wall before application. Brush the paint into the surface with a short-bristled brush. Apply first to the joints and then return to paint the entire wall. Because the goal is to coat the entire rough surface, use a scrubbing, not a painting motion. Allow 12 hours and recoat. Cure for 48 hours with adequate moisture and temperature. Portland cement paints containing latex need no moist curing.

OTHER PAINTS

Latex paints are made of polymer materials which are breathable, relatively odorless, easy to apply, and quick to dry. Latex paints are resistant to alkalies, but do not adhere well to chalky or glossy surfaces. Prepare the surface carefully by removing oil, scale, and chalk. Allow enough drying time (consult label) between coats.

Use a roller, spray, or tapered nylon-bristle brush to apply. The surface need not be dry. If surface is porous, or the weather is very dry, dampen the surface before painting. This is to allow the paint to flow better, not to help the paint cure. Drying is usually complete within one hour; moist curing is not required.

Oil-based paints and oil-alkyd paints are nonbreathing and subject to alkali damage, and thus are not used frequently with masonry. Apply these paints with a brush, roller, or spray and allow to dry per manufacturer's instructions between coats. When using epoxy or oil-based paints inside, make sure to provide adequate ventilation.

Do not apply in damp weather or at temperatures below 50°F. Apply with a wide brush, roller, or spray over a dry surface. Follow directions regarding drying time between coats; 24 to 48 hours may be needed.

Rubber-based paints offer great water resistance and are used for exterior and interior applications. The surface is nonbreathing and alkali-resistant. Rubber-based paints are suitable to interiors subject to water, such as bathrooms and kitchens.

Brush on two or three coats to a dry or slightly damp surface. Allow enough time to dry between coats (24 to 48 hours) or solvent in the second coat will degrade the first coat. Use with adequate ventilation and be cautious because the solvent is flammable.

SURFACE PREPARATION

Paint is only as good as the surface it is applied over, so surface preparation is an essential step in painting. For new surfaces, allow the masonry to dry and set for the specified period. For latex and portland cement paints, a few weeks is usually sufficient. But for paints that cannot be applied over damp or alkali surfaces, about six months is preferable.

All surfaces should be free of dust, grease, dirt, form-removing agents, and efflorescence. These contaminants can be removed by scrubbing, brushing, or air blasting. In extreme cases, sandblasting, pressure cleaning, or steam cleaning may be needed. Detergents such as TSP or caustic soda will remove grease and oil. Flush the surface thoroughly after cleaning and allow it to dry.

On porous surfaces, fill coats, sometimes known as primer-sealers or fillers, are brushed on to fill voids so the paint can form a smooth surface. Fill coats are often compounded of portland cement and silica sand. If latex admixtures are used, there is usually no need to moist-cure the fill coat.

Stir the paint before application, either with a power shaker, power stirrer, or a hand paddle. Do not power-mix latex paints—this will cause foaming.

APPLYING CLEAR COATINGS

Clear coatings are based on silicon and other chemicals. Detailed application instructions are provided by manufacturers. The following is a general guide to applying clear coatings:

1. Select the proper coating. Note the type of masonry surface and its orientation. (Some coatings are appropriate only to horizontal surfaces, others to vertical surfaces.) Limestone may need special material. Coverage per gallon may vary widely depending on coating material and surface porosity and texture. You may have to test to calculate material requirement.

2. Prepare the surface. Allow new work to age as the manufacturer specifies. Clean and repair old surfaces, and tuckpoint if needed.

3. Observe weather limitations during application.

4. Cover shrubbery and adjacent buildings.

5. Apply the coating with a low pressure spray or brush. Flood the surface so the material runs down the wall as far as indicated by the manufacturer.

6. Apply a second coat after the proper curing period.

APPLYING MEMBRANE WATERPROOFING

Panels of membrane waterproofing can be applied with adhesive to the outside of foundation walls. Some panels are manufactured with protective layers to prevent damage from backfilling and construction traffic. Follow the manufacturer's directions and observe the product's temperature limitations. The following procedure is a general technique for applying 4-foot × 8-foot panels.

1. Prepare the surface. The surface must be flat so the panel can bond thoroughly. Patch holes and voids. Parge block walls to smooth the surface.

2. Prime the entire surface with the correct primer at the suggested rate, using a roller. Two coats may be required over absorbent concrete.

3. Install the panels over the primer and roll under pressure to ensure adhesion. Protect the panels from construction traffic and other damage until backfilling is complete.

COLORING MORTAR

Mortar colors are used for esthetic reasons—to harmonize the mortar with, or contrast it to, the masonry units. Some jobs call for using two colored mortars at once. This can done by using two mortar tubs or dividing a single tub. Use separate trowels to avoid streaking one color into the other.

Colored mortar may be prepared 1) from bags of cement colored in the factory, or 2) by adding pigments to masonry cement at the job site. Mix either colored mortar until no streaks are visible. Sand color will affect the final mortar color, so use white sand for light-colored mortars.

Mortar colors are also influenced by handling and curing practices, weather conditions, admixtures, and the moisture content of the masonry units. Consistent measurement, mixing, and tooling practices will help ensure uniformity of color.

PIGMENTS

Pigments are best for small batches or when the necessary color is not available in prepared cement. Never use an organic coloring agent, as this will weaken the mortar. Do not use more than 10 percent pigment by weight of the cement in the mix (2 percent for carbon black).

Because mortar lightens as it sets and dries, the only sure way to achieve the desired color is to prepare a small batch and apply it to the masonry unit ahead of time. If this is not feasible, allow the sample to dry on an absorbent unit like a concrete block. This color will be close to that of fully set mortar.

In any case, use exact measurement of all ingredients to achieve consistent coloring. Test the mixing by flattening the mortar under a trowel. Any streaking indicates that more mixing is needed.

READY-MIXED

A variety of colors are available ready-mixed in masonry cement, for greater convenience, speed of mixing, and uniformity of color. The color selection is more limited than pigments, however. Tinted mortar is more expensive than standard mortars. Colored mortars are generally mixed in the same fashion as standard mortar, unless the manufacturer dictates otherwise.

COLUMNS

Columns are structures that support beams, joists, and other loads. Masonry columns are built of brick or block and are usually reinforced. Locate a column so the weight bears directly upon the column center. Industry standards call for the following relationships between length, width, and height:

- *Length shall be no more than three or four times width.*
- *Height shall be limited to twenty or twenty-five times width.*

Thus, using the maximum allowable dimension, a column 1 foot wide could be 4 feet long and 25 feet high. The exact dimensions depend on the application, the materials and reinforcing, and the load. Consult building codes and the architect if necessary.

A steel bearing plate is frequently used to spread the weight of building elements resting on columns. For maximum strength, the plate should bear on most of the column's cross-section. For information on a particular type of column, see *Piers,* p. 266.

BLOCK

Block columns may be laid with two U-shaped units per course or with single-core blocks, also called column blocks. Block columns are laid by combining techniques for columns, reinforced masonry, and grouting. Because columns often bear heavy loads, follow details on the plans carefully.

Use this procedure for laying reinforced block columns:

1. Lay out dry units to ensure proper bond.

2. Clean the foundation and begin laying the column in running bond. Lay each course completely and plumb it before laying the next course. Do not build up leads. Use full-bed mortaring unless plans specify otherwise.

3. Install reinforcing as specified in the courses and cavity. Grout with low- or high-lift procedures as specified.

4. For the top course, use solid units or grout the hollow units. If necessary, embed angle bolts in the proper positions. Or place the anchor plate in a bed of mortar (see Fig. 8.16).

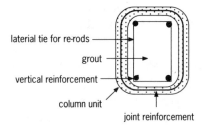

lateral tie for re-rods

grout

vertical reinforcement

column unit

joint reinforcement

Fig. 8.16 Reinforced block column

BRICK

Plain brick columns are laid up using the order marked in Fig. 8.17. All joints must be full. Loads can usually be placed on a brick column within two days of construction.

Brick columns can employ lateral and vertical reinforcement to increase strength. Follow practices for reinforced masonry, particularly regarding proper placement and overlap of reinforcement. Keep the grout space clean. Both low- and high-lift grouting can be used per specifications.

CONCRETE BLOCK CONSTRUCTION

Concrete block are used to build economical, durable, and strong structures. A great deal of engineering work has gone into designing both the units themselves and the techniques used for specifying and laying them properly.

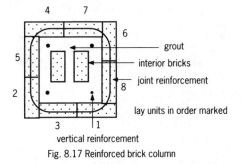

Fig. 8.17 Reinforced brick column

Units that are laid with some moisture in them will shrink after placement because of evaporation. To reduce shrinkage and bonding problems, blocks must be relatively dry when laid. Never wet block before use. Dry out block that was stored in humid conditions, or was rained upon, before use. Drying is also required for block that will be used for interiors where the relative humidity will be low (see *Construction practices, storing material*, p. 214).

Block can be air dried by stacking it loosely out of the rain, preferably in a sunny location. During cool or humid conditions, blow hot air from an oil or gas heater through the stack.

The foundation top should be clean to ensure that bed joint mortar will bond to it. Clean off laitance as well. Some specifications call for sandblasting or chipping to expose aggregate in the foundation for a good bond.

LAYING BLOCK

In many ways, laying block is similar to laying brick. Take care to fill all head and bed joints completely and avoid "crowding the line" by placing blocks too close to the line (see *Leads, building*, p. 217).

WHERE TO APPLY MORTAR

The face shells and webs of hollow blocks are thicker at the top than at the bottom. This is to provide a good surface to receive the bed joint mortar for the next course.

Block is generally laid with mortar on the face shells, not on the web. This is called face-shell bedding. Use full-mortar bedding—mortar on face shells and webs—in the following situations:

- *Columns, piers, and pilasters that will bear heavy loads.*
- *Concrete bricks and solid blocks.*
- *The first course for all blocks.*
- *Around cores that will be grouted, to contain the grout while it is liquid.*

Observe the following practices when laying concrete block:

- *In cold weather, wait longer for the mortar to set up before tooling the joints.*
- *In hot weather, do not to spread mortar too far ahead of yourself, to prevent mortar from setting up before you can lay the blocks.*
- *Keep mortar droppings out of cores that will be grouted to allow the grout to bond to the foundation.*
- *Do not reuse mortar that falls to the scaffold or ground because it will not have proper weather resistance.*
- *To avoid smearing the wall, wait until mortar drops are almost dry to cut them off. Smears are difficult to remove from blocks, and paint may not be able to hide them.*
- *Under either of the following conditions, you must remove and relay a block. Do this by removing the unit, scraping off the mortar, and replacing the unit on fresh mortar:*

  1. *Joints partly empty.* "Slushing" mortar into joint must be avoided because the resulting joint will not resist the weather.

  2. *Block in the wrong position or location.* Do not attempt to adjust blocks after mortar has stiffened or after the succeeding course has been laid. This will break the bond and lead to cracks and water penetration.

PLACING THE UNITS

After the leads have been built, or the masonry guides erected, stretch the line between the corners. Stand about three blocks on their ends, ready to receive head joint mortar. Mortar the face shells for the length of three blocks on the wall (usually 48 inches). Strive to get a full, even bead of mortar on both shells. Then mortar, or butter, the upper end of the standing three blocks to make the head joint. To ensure a full head joint, the best practice is to also butter the head joint of the block already laid.

Pick up one block and carefully lay it on the wall and shove it gently toward the adjacent block. Use the grip that is most comfortable and effective for you. Many masons place the unit by tipping it slightly toward themselves, allowing them to sight down the course below while placing the unit. Take care to place the block evenly, so it will extrude from both head and bed joints as you place the unit. Set the block so its top edge is level with the top of the line and about 1/32 inch away from it. Cut off mortar that extrudes from the joints and throw it back on the mortarboard or use it to butter the next block. Then pick up next block and repeat the procedure.

When all three blocks are placed, check again that the faces of the blocks are 1/32 inch from the line. Adjust if necessary by tapping the block with the trowel handle. Use the mason's level only to check that the block faces are flush with the rest of the wall. Continue with this process until you reach the opening for the last block in the wall. Now prepare the closure block by buttering all four of its head joints. Place the block, being careful not to push off too much mortar from the head joints.

Depending on weather and other conditions, tool the joints immediately after the course is laid or wait a few minutes. Perform the tooling when the mortar is hard enough to take an imprint from your finger but before it gets so stiff that it will crumble.

CONCRETE BRICK

Concrete bricks are modular units that are used by themselves or in combination with concrete block. Some building codes prohibit furrowing mortar for concrete bricks. Take care to adequately mortar head joints, and do not "slush" mortar into head joints that were not filled during laying. Relay the unit on fresh mortar.

BOND BEAM

Bond beams are reinforced concrete building elements that are cast in place to increase tensile and lateral strength in masonry walls.

The goal of bond beams is to prevent cracking and failure, and the technique can also make lintels or beams under the perimeter of walls. In areas subject to earthquake or high wind, more bond beams are used, and each beam may be larger as well.

Bond beams are made of grout and reinforcing rods that are held in bond beam blocks. Bond beams units fit the modular system. Some units have solid bottoms. Units with hollow bottoms must be laid on top of wire mesh to prevent the fresh grout from dropping through.

Bond beams are often found:

- *At the top of a story.*

- *Above an opening (to serve as a lintel).*

- *Above a floor to distribute the weight of the floor.*

- *Below a window sill.*

Locations and details of bond beams are found on specifications and blueprints. Look for the following information:

- *Size, spacing, and number of bars per beam.*

- *Orientation of the steel: horizontal, vertical, or diagonal.*

- *Thickness of the mortar joints around the bond beam.*

- *Type of grout to be used and quantity needed.*

(See *Reinforcing, placement*, p. 272).

Reinforcing bars are laid inside the units before grouting. For 8-inch walls, use at least two #4 bars. For 10-inch and 12-inch walls, use at least two #5 bars. Bars must bend around corners and T-intersections.

Bars must be centered in the bond beam and be completely surrounded by grout. Sometimes positioners are used to hold the rods in place during pouring, but the common practice is to use a hook to lift the rods immediately after the grout is poured.

For bond beams poured in tight cavities, it is important to prevent the steel from touching the masonry units or being exposed to air. Grout spaces must be larger than the reinforcement. The minimum size can be figured by 1) 1 inch or 1.33 times the size of the largest aggregate, whichever is larger, or 2) 1/2 inch or 1 inch larger than the sum of the diameters of the reinforcement. Most codes will allow smaller clearances in areas where the bars overlap each other.

Overlapping is used to join individual lengths of re-rod into longer beams. For strength, try to stagger the joints. Observe the proper overlapping distance, as given by number of bar diameters or inches of overlap. Thirty diameters is the general rule for overlap. Thus a #4 bar (1/2 inch) overlapped 30 diameters would need an overlap of 15 inches. Bars can also be joined by welding.

It is often possible to fill the bond beams while filling vertical cavities with grout. This saves time, but be careful to ensure that grout travels far enough horizontally to completely fill all bond beams (see Fig. 8.18).

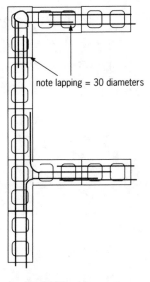

note lapping = 30 diameters

Fig. 8.18 Bond beam corner and tee-intersection

CORNERS

Corners must be laid to conform to the modular plan. 8-inch walls present no problem because the blocks are half-lapped at the corner. Walls of other thickness require adjustments to maintain the correct bonding. This can be done with L-shaped units or by cutting standard units to fit. Many corners require different layups on the alternate courses (see Fig. 8.19).

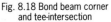

CONSTRUCTION PRACTICES

Good construction does not merely involve correct design and material selection. It also requires storing material properly, laying units and tooling joints correctly, and allowing mortar and grout to cure in conformance with industry standards. All masons should observe standard industry practices.

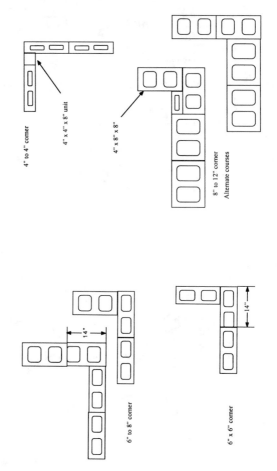

4" to 4" corner

4" x 4" x 8" unit

4" x 8" x 8"

8" to 12" corner
Alternate courses

6" to 8" corner

14"

6" x 6" corner

14"

Fig. 8.19 Block corner details

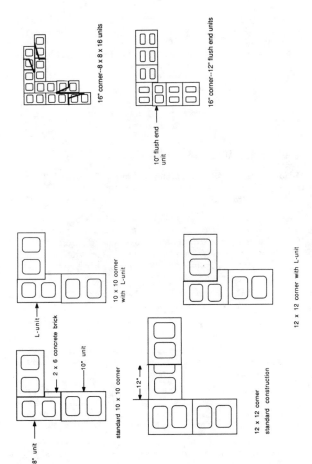

16" corner–8 x 8 x 16 units

16" corner––12" flush end units

10" flush end unit

10 x 10 corner with L-unit

standard 10 x 10 corner

L-unit

2 x 6 concrete brick

–10" unit

8" unit

12 x 12 corner with L-unit

12 x 12 corner standard construction

12"

Fig. 8.19 Block corner details (cont.)

STORING MATERIAL

All units, sand, and masonry cement should be kept dry before use. If rain or wet ground is expected, use pallets, 2 × 4 scraps, or other blocking to keep cubes of bricks and blocks off the ground. Plastic wrap is usually enough to keep rain off stored material, but be sure the plastic is weighted down so it cannot blow away. Store sand on a high spot so it cannot become saturated with runoff.

Mortar and portland cement are especially subject to water damage and must receive special attention at the job site. Store cement on pallets or other substantial blocking. In rainy areas, a double wrap of plastic may be advisable.

LAYING UNITS

The masonry industry has adopted certain trade practices to ensure that structures are strong, as watertight as possible, and attractive. These standards deal with mortar application, and laying and positioning units:

- *Read the blueprints and specifications to find the size and nature of the joints. For brick, bed joint thickness is usually specified in terms of a number of courses to a certain height, such as three courses to 8 inches. Block joints are usually 3/8 inch, to fit 4-inch or 8-inch modules. If possible, bed joints and head joints should have equal thickness.*

- *Joint thickness must be close to specification, although minimal variation may be needed to compensate for variations in unit size.*

- *Notice which joints (such as weeps, and collar and control joints) must remain free of mortar.*

- *Be aware of factors that may damage the bond, such as weather, mortar condition, foreign matter (including form release agents and dirt), and excess unit absorption.*

- *If the units have a suction greater than 30 grams per minute per 30 square inches, they must be wet prior to laying. Use a sprinkler or hose to wet the pile, and allow the surfaces to dry before laying. Wetting the day before is good practice (see Brick, absorbency, p. 47).*

- *Retemper hard-to-work mortar if it is less than about two hours old. Discard older mortar, unless you are using a retarder or working in cold weather.*

- *Joints must be filled with mortar. Voids can allow moisture to penetrate. Incomplete joints can reduce strength up to 60 percent and permit spalling and cracking due to freezing and thawing. Use enough mortar so you need not point up voids after laying. Do not slush mortar into head joints after laying, as this does not ensure a tight joint. Relay units that have defective mortar joints.*

- *If collar joints must be filled, do not use slushing, as it will not be effective. Fill the joint as you lay the second wythe.*

- *Some building codes prohibit furrowing, the lengthwise groove in bed joint mortar, because it can leave voids that permit water passage.*

- *Do not allow mortar to remain on the wall too long before placing units. The exact length of time depends on weather, unit absorption, joint thickness, and mortar type and condition. About one minute should be considered the maximum under normal conditions.*

- *Units must be adjusted and positioned while the mortar is still soft, and preferably by hand pressure alone. If units are moved after the initial set, they must be removed, cleaned, and replaced on fresh mortar. Otherwise, the joint will be likely to crack or leak.*

- *Mortar in the joints must be compressed while units are laid. If the weight of the unit does not provide enough compression, you must compress the mortar with manual force. (Uncompressed joints may develop cracks and are less resistant to water penetration.)*

TOOLING, CLEANING, AND CURING

The final steps in laying masonry are tooling, cleaning, and curing the mortar. Tooling is done for aesthetic and structural reasons, primarily to increase resistance to water in the joint. Cleaning is mainly for appearance. Curing is needed for mortar to achieve its full compressive and tensile strength. These trade practices will help insure sound and attractive construction:

- *Tool all joints at the same interval after laying to achieve consistent mortar color and proper weather resistance. Use a tooling method that forces the mortar into the joint and toward adjacent masonry surfaces. Tool the head joints first. For bed joints, run the jointer against the upper edge of units to ensure an attractive result. Joints that were not filled during laying must be raked, pointed, and tooled to match the surrounding area, preferably while the mortar is still green.*

- *Keep the structure as clean as possible during construction. Brush the joints when the mortar is set to remove burrs. Brushing too early will smear the wall.*

- *Cover the bottom 3 feet to 4 feet of wall to prevent spatter.*

- *Do not try to remove spatter immediately. After spatter has dried somewhat, remove it with a trowel. Use a chip of block or concrete brick to rub dry mortar from the face of concrete block.*

- *Masonry structures must be cured so as to maximize the hydration of the cement. Use special procedures during hot or cold weather to ensure proper hydration (see Weather, p. 297).*

- *Cover new construction to prevent rain from damaging the fresh mortar. Use a covering that will not blow away in the wind. Metal clamps can be used for this purpose.*

- *Clean masonry with the proper techniques and only to the extent necessary.*

COPINGS

Copings are structural elements used to prevent water from entering horizontal masonry elements, such as the tops of parapet walls. The design of copings involves a consideration of these factors:

- *Number, type, and size of joints in the coping.*
- *Differential movement between coping and wall.*
- *Type and location of flashing.*
- *Connection of the coping to the wall.*

Copings should overhang the wall and have at least 1 inch of drip to prevent water from running back on the underside of the projection and down the wall. Reducing the number of pieces will cut the chance of water entering the joints in the coping. Metal copings should be flexibly connected to allow differential thermal movement; copings of concrete, stone, or masonry can be fixed more tightly. Flashing is used under many copings to prevent water entry (see Walls, Parapet wall, p. 111).

CORNERS

Corners must carry the lead for straight walls. They are also vital visual elements of a building. Square corners are the most common type, but masons must also know how to build obtuse, acute, and radial corners.

Masons call corners "leads" when they are built first to provide guides for the intervening walls. Leads must be perfectly accurate, even if masonry guides are used. For long walls, an intermediate, or rake-back lead, is used to prevent the line from sagging.

The process of building leads is similar in brick and block work. For brick, about seven to nine courses are laid up as a lead. For block about three to five courses are laid. If too little lead is laid, the lead may not be set soon enough to support the line. Excessively tall leads reduce efficiency because they are, unit for unit, slower to lay than walls.

LAYING A SQUARE LEAD

Many varieties of square corners are possible, depending on bonds and materials (see *Bonds, Brick*, p. 175; & *Concrete Block Construction, Corners*, p. 211). Before starting the corner, mix mortar and stack units near the corner. Calculate how high you want the lead to go. A five-course lead requires three stretchers on one side and two on the other. A seven-course lead requires four stretchers on one side, and three on the other.

Use this procedure to lay out a lead in brick or block:

1. Mark the courses on the foundation by snapping a chalkline between batter boards or between nails placed in the foundation (see *Site Preparation and Layout*, p. 274).

2. In brickwork, lay out units dry to check the locations of corners and openings. In block work, use measurements instead. Adjust the spacing to reduce the number of cuts.

3. Mark the head joints on the foundation with a pencil.

4. Remove the dry units and lay out a bed of mortar for one tail of the lead. For unreinforced block construction, make this bed thicker than beds used for subsequent courses. For reinforced block construction, use a smaller bed joint to allow grout to contact the foundation for a good bond.

5. Lay the corner unit. Level and plumb the unit, making sure it conforms exactly to the line on the foundation.

6. Spread enough mortar for two more units in the tail you are laying. Level and plumb those units, then use the level as a straightedge to ensure all the blocks line up with the wall line. Check that head joints correspond to the pencil marks on the foundation.

7. Begin the second tail of the same course in the same fashion. Use a steel square to check that the corner is square. Lay the level across both tails to check that units are all level with each other. Check that the joint spacing is correct, using a brick spacing or modular rule.

8. Lay the remaining courses of the lead. Begin the process again at the other corner.

9. When completed, each unit in the lead must be level, plumb, at the proper height, and aligned with the wall. Bed joints must be of equal depth. Make sure not to disturb the leads while stretching the line between them (see Fig. 8.20).

OBTUSE AND ACUTE CORNERS

Angles other than 90° can be made in three ways: by using specially molded units, by sawing units to fit, and by lapping square-cornered units to form pigeonholes and dentils.

Molded units are the easiest way to form nonsquare corners if they are available and specified by the architect. On even-numbered courses, the molded unit is placed in one leg of the angle; it is reversed in odd-numbered courses.

Avoid using pieces 2 inches and smaller when cutting units for nonsquare angles, as these reduce the strength of the wall. Make sure to figure the cuts to avoid cores and cavities or use solid units if available.

Use the following procedure to cut bricks and solid blocks to form obtuse angles (a similar technique can be used to mark acute corner units) (see Fig. 8.21):

1. Lay out the walls with a chalk line from reference points.

2. Lay out wall (A) dry. Start the bond as specified in plans or use a convenient starting point, such as a square corner or large opening.

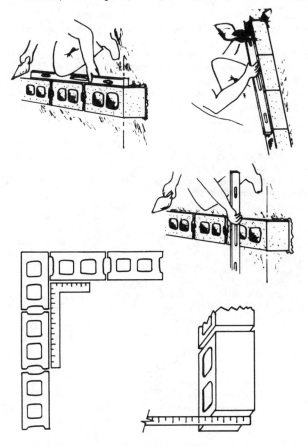

Fig. 8.20 Aligning leads

3. Place corner unit (A) to overlap wall line (B).
4. Snap wall line (B) across corner unit (A), or use a 4-foot level to mark the line. This is the cut line for corner unit (A).
5. Transfer angle (CDE) to a bevel square and scribe it on a scrap of wood for reference during cutting.
6. Cut corner unit (A). Abrade the cut surface with a brick or a block to improve its appearance if needed.
7. Repeat the process for corner unit (B). Remove corner unit (A) and place corner unit (B) dry so its corner is one head joint thickness from unit (A) at (F).
8. Mark the cut across unit (B) with chalk line or 4-foot level. Cut the unit, transfer the angle to a bevel, and mark it for future reference.
9. Mark the backing wythe if one is to be used. Place units, mark them, and cut to fit.
10. Lay out the first and second courses dry to check that your cuts and angles were correct.
11. Lay the wall following usual practices.

RADIAL

When viewed from the facing side, radial corners are termed convex (corner protrudes) or concave (corner indents). Start a radial corner

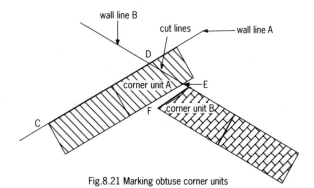

Fig.8.21 Marking obtuse corner units

by laying out the curve and sawing a plywood template to fit it, then use this template to align each course. Use several pieces of plywood if necessary for a large template (see Fig. 8.22).

LAYING OUT A CONVEX CORNER

The following procedure (see Fig. 8.22) will lay out a quarter-circle convex corner that will join smoothly with two walls oriented at right angles to each other:

1. Lay out wall lines (AB) and (BC).
2. Measure the length of one radius from (B) back to the start of the radial corner and mark point (D).
3. Repeat this procedure along wall (BC) and mark point (E). Draw square (BDFE). (F) is the radius point.
4. Draw arc (DE) on the foundation using (F) as the center and a string as long as the radius. Then lay the template plywood on the foundation and mark it.

LAYING OUT A CONCAVE CORNER

The following procedure (see Fig. 8.22) will lay out a quarter-circle concave corner that meets in a right angle with two walls oriented at 90° to each other:

1. Mark wall lines (AB) and (BC) to intersect at (B).
2. Measure one radius length from (B) along line (BA) and mark point (D), one end of the arc.
3. Do the same along line (BC) to mark the other end of the arc, point (E). Lengths (BD) and (BE) are both equal to the corner radius.
4. Mark the arc (DE) along the foundation using (B) as the center and a pencil tied to a string as long as the radius. Then lay the template plywood under the arc and mark again.

LAYING CONCAVE AND CONVEX CORNERS

Use the following procedure to lay a concave or convex corner:

1. Check the template for accuracy. Then lay it on the foundation and dry lay the first course. The narrower end of all head joints should be at least 1/4 inch wide. If the radius is short, saw units

rather than make an excessively wedge-shaped joint. (Molded units, with curved faces and angled ends, may be available. These units will make a smooth wall with rectangular, not wedge-shaped, head joints.)

2. Establish plumb points. Lay the template next to the arc on the foundation. Mark a plumb point about every two feet along the curve on the template and the foundation. These plumb points will enable you to replace the template in the same position while you lay the wall.

3. Lay the first course and check it with the template. Radial corners are usually laid one course at a time, with several bricklayers working together to hold the template and check alignment. Use a level to keep the bricks plumb.

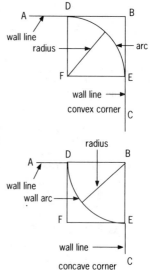

Fig. 8.22 Laying out radial corners

4. Check that the template is next to the plumb points on the wall, then bring all units into alignment. Continue laying the wall following accepted industry standards.

COURSES, SPACING

Course spacing is described in building specifications, and it is regulated by a story pole, a corner guide with factory-imprinted marks, or by a spacing rule. A brick spacing rule will regulate spacing in brick if the units are between 2-1/4 inches and 2-3/4 inches high.

A modular rule is used to regulate joint spacing for modular products.

A brick spacing rule can be complicated to read (see Fig. 8.23). Notice that the numbers are bunched together at the bottom of the rule and further apart at the top. This is because the difference in course spacing is very small when a single course is considered. But after a number of courses, this difference is multiplied and the marks are further apart.

Black numbers indicate the spacing number, with successive spacings 1/16 inch greater than the previous one. Note that spacing is given as 1 through 0, with 0 standing for 10. Thus, the smallest spacing is #1 and the largest #0 (10). Red numbers on spacing rules * indicate the number of courses at that height. Thus a line marked with a red 8 and a black 3 would indicate that this is the eighth course of a wall using spacing #3.

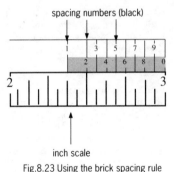

spacing numbers (black)

inch scale

Fig.8.23 Using the brick spacing rule

Follow this procedure to use a brick spacing rule:

1. Read the specification for course spacing, which is often expressed in terms of how many courses equals a certain height (for example, four courses to equal 11 inches).

2. Find that height on the inch side of the rule.

3. Reverse the rule (if necessary) and find the black number closest to the height (in this case, 6). This is the spacing number.

*Red numbers in fig. 8.23 are tinted gray here.

Measure from top of brick to top of brick, or from the top of the foundation to the top of any course.

4. Begin laying the brick, making sure that the top of every course remains even with a black 6 on the rule.

(For modular work, use a modular spacing rule. On this rule, the spacing indicates the number of courses that builds 16 inches of wall. Scale 2 on a modular rule is used for concrete block because it indicates that two courses equals 16 inches.)

Course height is also given in tables. These tables are handy but do not adapt to all the course spacings given on a spacing rule (see Table 8.4).

CUTTING AND SAWING

Hand cutting and sawing are used to shape or resize masonry units. For small jobs and soft materials and where accuracy is not important, hand cutting is adequate. But for greater speed, accuracy, and to cut hard and brittle materials, sawing is preferable.

HAND CUTTING

Hand cutting is done with a brick hammer, or a brick or block set struck with a mash hammer. Test the technique on a sample first. Support the masonry unit evenly—sand makes a good surface.

With experience, you can learn which cuts are easy, which are difficult, and which are impossible. Wear safety glasses or a face shield when cutting. Even though striking a chisel with a brick hammer is a common practice, for safety reasons, do not do so.

Two cutting techniques can be used for hand-held cuts: 1) Cutting away from you, which is done by holding the brick so the discarded end is away from you. 2) Cutting toward you is done so the discarded piece falls toward you. Cutting away is safer and preferable.

A description of cutting technique follows:

1. Set the unit on a uniform surface (or hold it in your hand if using a brick hammer).

2. Start the cut with soft blows, either with a set and mash hammer or the hooked end of the brick hammer. This will weaken the

TABLE 8.4
Height of Brick and Block Courses

Courses	Regular 4 2¼" bricks + equal joints =					Modular 3 bricks + 3 joints =	Concrete Blocks	
	10" ¼" joints	10½" ⅜" joints	11" ½" joints	11½" ⅝" joints	12" ¾" joints	8"	3⅝" blocks ⅜" joints	7⅝" blocks ⅜" joints
1	2 1/2"	2 5/8"	2 3/4"	2 7/8"	3"	2 11/16"	4"	8"
2	5"	5 1/4"	5 1/2"	5 3/4"	6"	5 5/16"	8"	1'4"
3	7 1/2"	7 7/8"	8 1/4"	8 5/8"	9"	8"	1'0"	2'0"
4	10"	10 1/2"	11"	11 1/2"	1'0"	10 11/16"	1'4"	2'8"
5	1'0 1/2"	1'1 1/8"	1'1 3/4"	1'2 3/8"	1'3"	1'1 5/16"	1'8"	3'4"
6	1'3"	1'3 3/4"	1'4 1/2"	1'5 1/4"	1'6"	1'4"	2'0"	4'0"
7	1'5 1/2"	1'6 3/8"	1'7 1/4"	1'8 1/8"	1'9"	1'6 11/16"	2'4"	4'8"
8	1'8"	1'9"	1'10"	1'11"	2'0"	1'9 5/16"	2'8"	5'4"
9	1'10 1/2"	1'11 5/8"	2'0 3/4"	2'1 7/8"	2'3"	2'0"	3'0"	6'0"
10	2'1"	2'2 1/4"	2'3 1/2"	2'4 3/4"	2'6"	2'2 11/16"	3'4"	6'8"
11	2'3 1/2"	2'4 7/8"	2'6 1/4"	2'7 5/8"	2'9"	2'5 5/16"	3'8"	7'4"
12	2'6"	2'7 1/2"	2'9"	2'10 1/2"	3'0"	2'8"	4'0"	8'0"
13	2'8 1/2"	2'10 1/8"	2'11 3/4"	3'1 3/8"	3'3"	2'10 11/16"	4'4"	8'8"
14	2'11"	3'0 3/4"	3'2 1/2"	3'4 1/4"	3'6"	3'1 5/16"	4'8"	9'4"
15	3'1 1/2"	3'3 3/8"	3'5 1/4"	3'7 1/8"	3'9"	3'4"	5'0"	10'0"
16	3'4"	3'6"	3'8"	3'10"	4'0"	3'6 11/16"	5'4"	10'8"
17	3'6 1/2"	3'8 5/8"	3'10 3/4"	4'0 7/8"	4'3"	3'9 5/16"	5'8"	11'4"
18	3'9"	3'11 1/4"	4'1 1/2"	4'3 3/4"	4'6"	4'0"	6'0"	12'0"

TABLE 8.4 (cont.)
Height of Brick and Block Courses

Courses	Regular 4 2¼" bricks + equal joints =					Modular 3 bricks + 3 joints =	Concrete Blocks	
	10" ¼" joints	10½" ⅜" joints	11" ½" joints	11½" ⅝" joints	12" ¾" joints	8"	3⅝" blocks ⅜" joints	7⅝" blocks ⅜" joints
19	3'11 1/2"	4'1 7/8"	4'4 1/4"	4'6 5/8"	4'9"	4'2 11/16"	6'4"	12'8"
20	4'2"	4'4 1/2"	4'7"	4'9 1/2"	5'0"	4'5 5/16"	6'8"	13'4"
21	4'4 1/2"	4'7 1/8"	4'9 3/4"	5'0 3/8"	5'3"	4'8"	7'0"	14'0"
22	4'7"	4'9 3/4"	5'0 1/2"	5'3 1/4"	5'6"	4'10 11/16"	7'4"	14'8"
23	4'9 1/2"	5'0 3/8"	5'3 1/4"	5'6 1/8"	5'9"	5'1 5/16"	7'8"	15'4"
24	5'0"	5'3"	5'6"	5'9"	6'0"	5'4"	8'0"	16'0"
25	5'2 1/2"	5'5 3/8"	5'8 3/4"	5'11 7/8"	6'3	5'6 11/16"	8'4"	16'8"
26	5'5"	5'8 1/4"	5'11 1/2"	6'2 3/4"	6'6"	5'9 5/16"	8'8"	17'4"
27	5'7 1/2"	5'10 7/8"	6'2 1/4"	6'5 5/8"	6'9"	6'0"	9'0"	18'0"
28	5'10"	6'1 1/2"	6'5"	6'8 1/2"	7'0"	6'2 11/16"	9'4"	18'8"
29	6'0 1/2"	6'4 1/8"	6'7 3/4"	6'11 3/8"	7'3"	6'5 5/16"	9'8"	19'4"
30	6'3"	6'6 3/4"	6'10 1/2"	7'2 1/4"	7'6"	6'8"	10'0"	20'0"
31	6'5 1/2"	6'9 3/8"	7'1 1/4"	7'5 1/8"	7'9"	6'10 11/16"	10'4"	20'8"
32	6'8"	7'0"	7'4"	7'8"	8'0"	7'1 5/16"	10'8"	21'4"
33	6'10 1/2"	7'2 5/8"	7'6 3/4"	7'10 7/8"	8'3"	7'4"	11'0"	22'0"
34	7'1"	7'5 1/4"	7'9 1/2"	8'1 3/4"	8'6"	7'6 11/16"	11'4"	22'8"
35	7'3 1/2"	7'7 7/8"	8'0 1/4"	8'4 5/8"	8'9"	7'9 5/16"	11'8"	23'4"

TABLE 8.4 (cont.)
Height of Brick and Block Courses

Courses	Regular 4 2¼" bricks + equal joints =					Modular 3 bricks + 3 joints =	Concrete Blocks	
	10" 1/4" joints	10½" 3/8" joints	11" 1/2" joints	11½" 5/8" joints	12" 3/4" joints	8"	3⅝" blocks 3/8" joints	7⅝" blocks 3/8" joints
36	7'6"	7'10½"	8'3"	8'7½"	9'0"	8'0"	12'0"	24'0"
37	7'8½"	8'1⅛"	8'5¾"	8'10⅜"	9'3"	8'2¹¹⁄₁₆"	12'4"	24'8"
38	7'11"	8'3¾"	8'8½"	9'1¼"	9'6"	8'5⁵⁄₁₆"	12'8"	25'4"
39	8'1½"	8'6⅜"	8'11¼"	9'4⅛"	9'9"	8'8"	13'0"	26'0"
40	8'4"	8'9"	9'2"	9'7"	10'0"	8'10¹¹⁄₁₆"	13'4"	26'8"
41	8'6½"	8'11⅝"	9'4¾"	9'9⅞"	10'3"	9'1⁵⁄₁₆"	13'8"	27'4"
42	8'9"	9'2¼"	9'7½"	10'0¾"	10'6"	9'4"	14'0"	28'0"
43	8'11½"	9'4⅞"	9'10¼"	10'3⅝"	10'9"	9'6¹¹⁄₁₆"	14'4"	28'8"
44	9'2"	9'7½"	10'1"	10'6½"	11'0"	9'9⁵⁄₁₆"	14'8"	29'4"
45	9'4½"	9'10⅛"	10'3¾"	10'9⅜"	11'3"	10'0"	15'0"	30'0"
46	9'7"	10'0¾"	10'6½"	11'0¼"	11'6"	10'2¹¹⁄₁₆"	15'4"	30'8"
47	9'9½"	10'3⅜"	10'9¼"	11'3⅛"	11'9"	10'5⁵⁄₁₆"	15'8"	31'4"
48	10'0"	10'6"	11'0"	11'6"	12'0"	10'8"	16'0"	32'0"
49	10'2½"	10'8⅝"	11'2¾"	11'8⅞"	12'3"	10'10¹¹⁄₁₆"	16'4"	32'8"
50	10'5"	10'11¼"	11'5½"	11'11¾"	12'6"	11'1⁵⁄₁₆"	16'8"	33'4"

COURTESY OF THE INDIANA LIMESTONE INSTITUTE OF AMERICA

unit at the desired location. For difficult cuts, such as to make a split, many soft blows are needed.

3. If using a brick hammer, reverse it and use the square face to strike harder blows to finish the job; or increase the power of your swings on the mash hammer.

4. Use the hooked end of a brick hammer to clean rough spots after the break.

As an alternative, for small cuts, such as to dress a brick or stone before placement, strike the unit with the trowel heel. Hold the trowel sideways, with the bottom of the blade facing the hand holding the unit.

Stone is cut in much the same way, except that more work is usually required. Carefully assess the grain of the stone before beginning. Many stones have natural cleavages which permit simple breaks along the cleavages but prevent breaks in other planes.

SAWING

Always wear safety glasses when sawing. Use a mask. Use a respirator when sawing with a dry blade. Check that the saw is safely set up (see *Tools, saw*, p. 25). For economy of labor, cut all units of one size at one time. Place units to be cut in a handy location. Keep the carriage clean so units will sit squarely and not rock or shift while working. Use this procedure:

1. Set the adjustable guide on the saw carriage to the proper size and angle. Apply gentle pressure to the swinging arbor with your foot (for a pedal-operated saw) or your hand.

2. Score the unit about 1/16 inch deep along the cut line.

3. Make four or five passes to complete the cut, always holding the unit tight to the carriage and using steady pressure on the cutting wheel.

DOORS

Masons deal with doors by 1) leaving a rough opening for later installation of the frame and door, or 2) installing the door frame during construction. Doors are described and specified on the blueprints, specifications, and door schedule. Consult all three to find the dimensions, location, and other details on doors.

ROUGH OPENING

Masons commonly create a rough opening in the wall, which is then fitted with a door frame to make the finished opening. Much work can be saved if the door frame conforms to modular dimensions. The rough opening is figured from the bottom of the sill to the bottom of the lintel. When planning the opening, choose a course height that will align the top of a course with the top of the opening.

Opening locations are often described with a centerline on blueprints. Mark this centerline on the foundation, then measure one-half the rough opening width in each direction, and lay out the bond for the first course. Plumb the jamb while laying the wall. Make sure that half the units at the jamb have smooth faces. Set any required anchors as you proceed with laying the door. Consult the manufacturer's directions regarding anchors.

FINISHED

You can also install the door frame and build the wall around it. Plan the wall as above. If a lug sill is specified, set it into place and allow it to set before laying the wall. (A slip sill can be installed after laying the wall.)

Rest the frame on the lug sill or foundation and brace it into place, ensuring that it remains plumb, square, and correctly located. Use a spreader to keep the side jambs parallel during construction. If the opening is wide and the lintel will be cast in place, add vertical braces as necessary to prevent sagging in the middle of the head jamb. (These braces should not be needed when using a steel or precast lintel.) Make a final check that the jamb is correctly positioned and adjust braces as needed. During layup, anchor the frame to the jambs as specified (see Fig. 8.24).

METAL FRAMES

Metal door frames come in two varieties: butted and wrap-around. Butted frames are no deeper than the thickness of the wall. Wrap-around frames are slightly deeper and extend about 2 inches past the

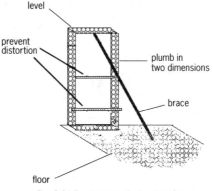

level

prevent
distortion

plumb in
two dimensions

brace

floor

Fig. 8.24 Bracing and aligning door frame

jamb. Both types of frames are assembled on the job site and placed, aligned and braced in the correct location. Check that anchors fit into coursing before installing.

Metal frames may be placed before or after the wall is built, depending on specifications and frame type.

SILL

Two types of sills are common: lug and slip. Lug sills must be installed before the wall is built because they have 4-inch surfaces at their outer edges that must underlie the jamb units. Lug sills must be levelled across the pads to ensure proper drainage. The lug sill is bonded into place with mortar at the ends, but the bed joint under the opening is left empty until the wall is completed. Then this joint is filled by tuckpointing. This procedure prevents cracking if minor settlement occurs during construction.

Slip sills can be placed after the frame is installed. Bond the sill into position, level and plumb. Then check the overall size of the opening to make sure it conforms to specifications. After the mortar under the sill has set, rake bed and head joints and apply caulking in accordance with the manufacturer's instructions.

FLASHING

Flashing has two basic functions in masonry structures: 1) to prevent water from entering, and 2) to divert water that has entered to the outside. Flashing seals masonry to itself or other building material. For large jobs, flashing is specified on blueprints and it may be fabricated in the shop. For smaller jobs, the masons design and construct the flashing.

Common locations for flashing are the junction between chimney and roof, at the base of a wall, beneath sills and copings, above lintels, and level with floors.

Flashing at the base of a veneer wall must extend 8 inches vertically up the backing wall. Through-wall flashing, used for solid masonry walls, is inserted into the backup wythe above its entry into the face wythe. Install weepholes in the facing wythe every 16 inches to 24 inches O.C. and just above the flashing, to allow water to escape the wall.

Flashing provides a substantial bond break in a wall. If mortar is placed above and below the flashing material, the flexural strength of the wall may be reduced as much as 70 percent. This is desired in some locations, as where differential movement is expected (see Fig. 8.25).

Flashing beneath coping prevents moisture that leaked under the coping from penetrating the wall. This flashing must extend the full length of the wall (see *Walls, veneer, parapet and cavity*, p. 125; *Copings*, p. 216).

Flashing must be overlapped at horizontal joints at least 6 inches. The lap should be sealed with the proper mastic or adhesive. Flashing that does not extend the entire length of a wall, such as under a window sill, needs an end dam to prevent water from spilling off the end of the flashing. Construct such a dam by folding the end of the flashing up into a head joint.

In general, flashing should extend through the facing wythe and have a drip to keep water from running down the wall. Flashing should be continuous around corners, which requires the same lapping as other horizontal joints between flashing.

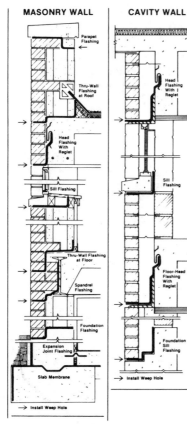

MASONRY WALL

Parapet Flashing

Thru-Wall Flashing at Roof

Head Flashing With Reglet

Sill Flashing

Thru-Wall Flashing at Floor

Spandrel Flashing

Foundation Flashing

Expansion Joint Flashing

Slab Membrane

Install Weep Hole

CAVITY WALL

Head Flashing With I Reglet

Sill Flashing

Floor-Head Flashing With Reglet

Foundation Sill Flashing

Install Weep Hole

VENEER

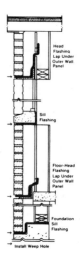

Head Flashing Lap Under Outer Wall Panel

Sill Flashing

Floor-Head Flashing Lap Under Outer Wall Panel

Foundation Sill Flashing

Install Weep Hole

Courtesy of Wasco York Flashings

Fig. 8.25 Flashing for solid, cavity and veneer walls

FOOTINGS

Footings are concrete pads which rest on earth or fill to support the entire structure. The person who coined the phrase, "An ounce of prevention is worth a pound of cure" might well have been looking at a masonry building that cracked because the footing settled. Good footings are not expensive, but the consequences of poor footings can be. On large jobs, the footings are installed by the concrete contractor; on small jobs the masons may be responsible for them. Some footings are poured at the same time as foundation walls; but more often they are poured separately.

The first step is deciding how large to make a footing. Building codes regulate footing size and construction, and they should also be described on foundation drawings. Eight inches is the usual minimum thickness of footings unless blueprints specify otherwise. For residential walls, the rule of thumb is to make the footing as deep as the wall is thick and twice as wide. Thus for an 8-inch wall, the footing would be 16 inches wide and 8 inches thick. This formula, however, is only useful when the underlying soil has good bearing strength and relatively low walls are being built. The Portland Cement Association recommends the following bearing capacities for soils:

LOAD CAPACITIES OF SOILS

Soil type	Tons per Square Foot
Soft clay	1
Firm clay or fine sand	2
Compact fine or loose, coarse sand	3
Loose gravel or compact coarse sand	4
Compact sand–gravel mixture	6
Solid rock	Practically unlimited

Soil does not always fall into such neat classifications, so there may be some variation between the book values and actual conditions. If in doubt, consult an engineer or architect.

In any case, the footing should extend below the frost line, and be at least 18 inches below grade.

To avoid different rates of settlement between a fireplace and the building, calculate the footing size so the bearing pressure of the fireplace roughly equals that of the building. Fireplace footings are often poured 12 inches thick and reinforced with bars. Blueprints should specify all facets of fireplace footing design.

If excess soil was excavated, fill the voids with gravel or concrete before pouring the footing. Never backfill with dirt. If some soil beneath footings was disturbed, compact it before pouring concrete. For small jobs, a hand tamper can be used, but a power compactor is better for large jobs.

Footings should have plumb walls and level bottoms to transfer force evenly to the soil. Wall footings can be poured into a trench. If the trench is smooth, no forms are necessary, but if the trench is rough or specifications require it, forms must be used. Forms can be simple as long as they offer enough support to hold the fresh concrete and their tops are at the proper height and level.

Generally, footings are not reinforced unless this is required by codes, blueprints, or conditions. If rods are needed, make sure they do not rest on the ground. Rods should be held at the correct height with crutches during pouring.

Several styles of footings are in use. Choosing the style is partly a matter of personal preference, and partly a matter of application. An optional key increases the grip between wall and footing; this can also be achieved by embedding re-rod dowels in the footing and extending them into the wall. Staggered or tapered footings use relatively little concrete for the amount of bearing surface they provide.

Stepped footings are used for sloping grades. For footings under block foundation walls, make the step dimensions modular to conform with block course height and length.

INSTALLING DRAIN TILE

Drain tile is often used to remove water from the vicinity of footings and foundations. In some areas, drain tile is required by code; in other areas it is required by common sense, and in still others, it is unnecessary.

The pipe comes in various materials, diameters and lengths:

- *four-inch or six-inch diameter concrete or clay tile, in 1-foot sections.*
- *perforated PVC or composition pipe, in 10-foot sections.*
- *perforated plastic pipe, in continuous coils.*

Ten-foot PVC pipes are excellent for drain piping because their smooth interior promotes water flow. Continuous plastic piping is convenient to lay, but you must take care to maintain the proper pitch.

Lay the pipe on the outside of the footing on a bed of stone filler material. For pipe that is made in 1-foot sections, wrap each joint with building felt to prevent soil from entering. Cover the pipe with at least 2 feet of 3/4-inch to 1-1/2-inch washed stone. Cover the stone with a layer of filter matting, straw, or building felt, to prevent fine material from plugging the tile. The filter matting can be draped from the top of the wall and over the washed stone for maximum effectiveness.

Pitch the pipes so the water will flow to a bleeder pipe that runs through the footing (bleeders are often located on 20-foot centers). The bleeder pipes connect to a sump, from which the water is piped to a storm sewer or other discharge. If the water will be pumped to a surface discharge, place the discharge far enough from the building that water cannot run back to the foundation. Outside, above-grade piping must be pitched steeply enough to prevent freezing in winter.

SUPPLIES AND EQUIPMENT FOR FOOTINGS

Use the following information for preparing bids and material lists for footing work:

Information and preparation:
Blueprint and specifications
Building permit
Check sewer, underground utilities, and water table
Excavation
Access to job and scheduling
Layout
Location of wall, chimney, column, and pilaster footings

Materials:
Concrete
Reinforcing bar and mesh
Drain tile
Form materials, stakes, and nails
Filter matting

Material for sump
Calcium chloride if specified
Sand and gravel fill

Equipment:
Steel forms
Water level or transit
Water pump if needed
Hay, blanket, plastic, or canvas coverings as needed
Pouring chute
Planks for wheeling and pouring concrete

GLASS BLOCK CONSTRUCTION

Glass block is laid much like other masonry units, except that it is not a structural element and needs more reinforcement. A glass block structure must have room for movement because the material expands at a different rate than masonry. The usual practice is to rest the sill blocks on asphalt emulsion and hold jamb and top blocks inside special channels. Expansion strips and caulking are used to provide weatherproof seals at the edges of panels and at movement joints.

Pittsburg Corning Corporation recommends that glass block be laid with Type S mortar and clean, white quartzite sand. The sand should be essentially free of iron, which can stain the surface of solar reflective glass blocks. For exterior use, add an integral waterproofing agent (such as a stearate compound) to the mortar.

Use this procedure (see Fig. 8.26) for laying glass block:

1. Install the jambs.

2. Lay out the wall dry and mark the bed joints on the foundation (marks should be visible after coating with asphalt).

3. Coat the sill with a heavy layer of asphalt emulsion and allow at least 2 hours for drying.

4. Affix expansion strips to the head and jambs. Make sure the strips cover the entire opening except the sill.

5. Using a stiff mortar, lay a full bed on the sill. Do not furrow the mortar. Place the first course of block with a joint thickness ranging from 1/8 inch to 3/8 inch. Adjust blocks by tapping

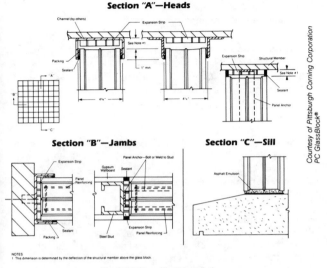

Section "A"—Heads

Section "B"—Jambs

Section "C"—Sill

*Courtesy of Pittsburgh Corning Corporation
PC GlassBlock®*

NOTES
1 This dimension is determined by the deflection of the structural member above the glass block.

Fig. 8.26 Glass block installation

with a rubber or wood tool, not a steel one. Use wedges if
needed to prevent blocks from squeezing out too much mortar.

6. Use joint reinforcing 24 inches O.C. vertically for blocks with
    3-7/8-inch thickness. Reinforce joints immediately above and
    below all openings. Reinforcing must extend the full width of
    the panel. Use the following procedure for installing
    reinforcing:

    a. Lay out one-half of the mortar. Do not furrow.
    b. Press the reinforcing into the mortar.
    c. Apply the second half of the mortar, without furrowing,
       and lay the joint.

7. Continue laying the blocks up to wall height. Keep mortar out
    of the spaces in the head and side jambs.

8. Strike joints while the mortar is still plastic. Remove mortar
    from block faces. Wipe extra mortar away with a heavy sponge,

but avoid scratching block faces. Rake joints that will be caulked to the required depth.

9. Allow the panel to set for 24 hours (longer in cold weather) and install packing into head and jambs that were not mortared. Then caulk the head and jamb. Seal other joints as required.

10. Clean the block faces. Do not use harsh cleaners, steel wool, or wire brushes. Wash away extra sealant and caulking with the proper solvents. Use sponge and water to remove mortar film as needed. Use caution so aggregate does not scratch the surface.

The advent of glass end blocks has allowed wider use of glass block. Special procedures are needed to ensure that end blocks are fully supported. Walls must be supported on three sides, and joint reinforcement must extend to within 2 inches to 4 inches of the outside edge of each end block.

End blocks have the finished side to the side in walls that have one side open. This is called "end condition." End blocks have the finished side across the top of a wall that is in "head condition."

Use this procedure for laying glass end block in end condition, in which the wall is supported on top, bottom, and one side:

1. Follow the general specifications for laying glass block described above.

2. Lay the first course, starting at the jamb and proceeding to the exposed end.

3. Lay joint reinforcement on top of the first course running from the side jamb to within 2 inches to 4 inches of the outside edge of the end block.

4. Install horizontal reinforcing in alternate courses and below the final course. Each horizontal reinforcing must reach within 2 inches to 4 inches of the end of the end block.

Use this procedure (see Fig. 8.27) for laying glass end block for a wall in head condition, in which the wall is supported on the bottom and two sides:

1. Follow general specifications for laying glass block as described above.

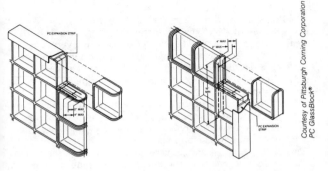

a) end condition

b) head condition

Fig. 8.27 Glass end blocks

2. Lay the first course.

3. Lay joint reinforcement for the full width of the panel.

4. Install horizontal reinforcing in every third course until the second-to-last course.

5. Lay the second-to-last course with vertical pieces of reinforcement in every second head joint, and the first and last head joints. Each vertical piece must terminate within 2 inches of the top of the wall. One cross wire must reach within 4 inches of the wall top.

6. Bend the last piece of horizontal reinforcement and slip it over the vertical reinforcements. Lay the final course, making sure not to disturb the existing blocks and reinforcement.

GROUTING

Grouting is the process of filling voids with a soupy concrete mixture known as grout. The goal is to stiffen the structure and bond elements together in bond beams, collar joints, and vertical cores. Grout is also used to secure anchors and firm up columns and piers. Building codes often specify locations of grouted cores and bond beams, and they may also require use of a certain grade of mortar or concrete in the grout.

Grout is always used in reinforced masonry construction to form bond beams and vertical reinforcing elements. Grout is also used in some unreinforced structures, such as in loadbearing walls, where it increases compressive and shear strength and reduces sound transmission. Full grouting is used for security reasons in prisons and banks. Sand can replace grout to reduce sound transmission through interior walls of unreinforced construction.

Bond beams are often filled when grouted cores are poured. If grouted cores are more than about 4 feet apart, you may need to pour the bond beams separately because grout may not flow far enough to fill the beams.

Quantities of grout required vary depending on how many cores and bond beams are being filled, and whether the collar joint or other areas will be grouted (see *Estimating Jobs*, p. 309).

A lift is a single filling of grout. Several lifts performed in a short time, before the grout sets, are termed a pour. Grouting is either low-lift or high-lift. Low-lift grouting, in which a few feet are poured at one time, is simpler and more commonly done, but high-lift grouting can save time and labor. In high-lift grouting, as much as one floor's height of wall is poured at one time, although usually in several lifts.

Low-lift grouting should be done as soon as possible after a section of wall is laid up. This practice allows all materials to cure at similar rates, reducing shrinkage cracking. Grout in block cores can be placed within 24 hours of laying the wall. Grout poured in collar joints may tend to explode the wall, so the wall should cure at least three days in warm weather and five days in cold weather before pouring.

After placement, the grout is immediately vibrated into place or rodded with a 1 × 2 stick to fill the voids. Grout with the proper consistency will need little agitation at this stage. If the units are absorbent, return after about 15 minutes to rod or vibrate again. This will prevent shrinkage cracking in the grout. Avoid overvibrating, which can damage units and reinforcement, or cause blowouts or separation of the grout from the reinforcement. When work stops for more than one hour, leave a key for the next pour by stopping the pour 1 inch below the block top. This key (required by many building codes) allows the next lift to lock to the previous ones.

Grout generally contains enough water to cure successfully. In areas that are especially hot, dry, and windy, it may be necessary to wet the outside of the structure to aid curing. In cold weather, follow practices for cold-weather masonry.

PREPARING

Make sure to provide enough clearance for grouting in vertical columns. Two-core block is better than three-core because it offers a larger cavity and is thus easier to fill. Cores must have minimum dimension of 2 inches in each direction. The cavities must line up vertically and mortar must not extrude from bed joints more than about 3/8 inch. While laying block, you can control extrusions by not placing mortar closer than about 1/4 inch to the inside of bed joints near grouted cores. Slope the mortar away from the core before placing the unit.

Block webs surrounding vertical cores that will receive grout need not be mortared if 1) all cores will be filled, or 2) flush end units are used at cores to be grouted. If not all cores will be grouted and standard units are being used, mortar the web joints around grouted cores to prevent grout from escaping.

If long pieces of vertical reinforcement are placed before the wall is laid up, use special A- or H-shaped units, which are easier to place around the existing reinforcement. If these units are not available, two-core units are better than three-core units because they fit more easily over the rods.

Install wire spacers to ensure that both horizontal and vertical reinforcing rods are held in the correct location in the grout during pouring. Cleanouts used in high-lift grouting are handy if you need to secure the bottom of vertical steel. Make sure to overlap rods the correct distance—generally 30 diameters. A common requirement is to secure bond beam rods within the beam every 192 diameters.

Install a vertical barrier in the grout space at intervals no more than 26 feet to prevent excessive horizontal travel of the grout.

Bond beam units have either solid or hollow bottoms. If the wall is not to be solidly grouted, install wire mesh under hollow-bottom blocks to hold the wet grout. Some units have knockouts that must be removed before inserting the rods. At corners and intersections, you may have to saw notches to allow room for the re-rod.

Grout space. If the grout space is greater than 2-1/2 inches, pea gravel can be used; otherwise a finer grout should be used. In any case, the minimum cavity dimension should be 1 inch. When the smallest dimension of a void is less than 2 inches, the maximum height of a pour should be 12 inches, according to the ANSI Building Code Requirements for Reinforced Masonry.

Pouring. Grout is generally fed by a ready-mix truck into a pump which distributes it to the work site. On smaller jobs, grout may be placed into cores using special buckets with funnels. The operation should be done as rapidly as possible to preserve the concrete's strength.

Use care when pouring grout, as it can stain the facing. Some masons prefer to work from the inside to minimize staining. Do not spill grout on the top of the wythes, as this can interfere with mortar bonding. Spread out your work along the wall to prevent blowouts from the hydrostatic pressure of fresh grout. A wall that does blow out must be torn down and rebuilt, even if it only moves a little bit.

Leave the top 1 inch of the cavity empty to allow the next lift to "key" to the previous lift.

LOW LIFT

In low-lift grouting, grout is pumped into cores after no more than 5 feet of wall is finished. Bond beams may be grouted at the same time as the cores. This technique requires no special equipment or masonry units and is necessary, for example, when window openings prevent high-lift grouting. Only 30 diameters of vertical reinforcing rod must extend from the cavity top (in high-lift grouting long sections of vertical re-rod can interfere with placing units). With each pour, another rod can placed into the cavity to form a continuous tie. In collar joints, the height of each lift should be no more than six times the collar joint depth, or a maximum of 8 inches to 12 inches.

HIGH LIFT

High-lift grouting is usually done after each full story has been laid up. High-lift grouting is more complicated than low-lift grouting and cannot be used where window openings, reinforcement, or other obstructions would prevent grout from flowing into cores and bond beams. Where it can be used, high-lift grouting saves time because fewer work breaks are needed. Depending on conditions, it is possible to pour up to 16 feet of 8-inch single-wythe wall in a single pour (though several lifts—commonly 4 feet each—are required for this pour).

High-lift grouting requires better clearance in the cores to ensure that the grout fills them. The minimum dimension of cores is 3 inches and the minimum area is 10 square inches. If the collar joint is to be grouted, it must be between about 2 inches and 6 inches wide (check local codes). Use special care when grouting collar joints because of the danger of bursting the wall.

If the collar joint is to be grouted, wall ties should not contain drips, which reduce their ability to withstand the hydrostatic pressure. The facing wythes must be of proper material to bond to the grout. Bricks should be of medium absorption and not have die skin, glazed, or heavily sanded surfaces.

Do not apply mortar near the inside edge of cores that will be grouted, to prevent extrusions from blocking the cores. Clean the cores twice a day with a long stick to knock off mortar protrusions.

Cleanouts at least 3 inches × 4 inches must be provided at the bottom of every core containing vertical reinforcement that will be grouted. The cleanouts allow removal of all mortar droppings and other debris. In brick wythes, cleanouts can be formed by leaving every third brick dry on the bottom course. Cover the foundation of the cleanout core with sand or plastic to prevent mortar from sticking to it. Before grouting, scrape or blow out the debris. Then insert the unit that was omitted from the wall or place a heavy weight against the cleanout opening to ensure that grout does not leak from it. Consult building codes for details on cleanouts.

Pouring must wait until the mortar has set sufficiently. Consult local codes. Pouring is done in successive lifts to reduce hydrostatic pressure on the bottom units and to allow some shrinkage and settling between pours. Puddle or vibrate grout within about 10 minutes of pouring. The next lift should be performed at 30 to 60 minutes later, after the previous lift has taken its initial set. Then, puddle again, reaching the tool down 12 inches to 18 inches into the preceding lift to force it tight against the walls (some shrinkage drying will already have occurred). Make sure to puddle the top lift after the waiting period to refill any spaces left by shrinkage.

MOVEMENT JOINT

Movement joints are used to allow adjacent sections of a building to move slightly with respect to one another. They alleviate cracking caused by differential movement, which results from changes in temperature, water content, and foundation shifting. In effect, the movement joint creates a planned, built-in crack that allows this movement to take place without causing harm. Movement joints offer a "path of least resistance" to the structure (see *Movement, differential*, p. 262). Other techniques used to prevent unplanned cracking are proper design, materials, workmanship, and joint reinforcement.

Movement joints are either control joints (used to allow shrinkage and expansion in concrete masonry) or expansion joints (used to allow expansion in brick masonry (see *Materials, Movement joints*, p. 67). Movement joints can be constructed on the job or installed ready-made.

A properly designed and installed movement joint provides all the advantages of a masonry wall while preventing damage from differential movement.

Movement joints have three functions:

- *Allow for expansion and contraction.*
- *Structurally join adjacent sections of wall.*
- *Seal against vision, sound, and weather.*

Movement joints have five elements: 1) The standard or special masonry units forming the sides of the joint. 2) The opening that provides room for expansion and contraction. 3) The provisions for lateral support. 4) The bond break that allows adjacent sections to shift relative to each other. 5) The seal that prevents water entry and provides an opaque appearance.

Unless the architect specifies otherwise, joint reinforcement must not extend through a movement joint, as this would prevent the desired movement. The general practice is to terminate horizontal reinforcement about 2 inches from the edge of the joint.

LOCATING MOVEMENT JOINTS

Movement joints should be located where differential movement is most likely to crack a building. Otherwise, stresses can cause cracks elsewhere, defeating the purpose of the joint. Movement joint styles and locations are specified in blueprints (see Fig. 8.28).

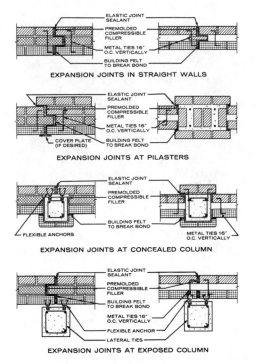

EXPANSION JOINTS IN STRAIGHT WALLS

EXPANSION JOINTS AT PILASTERS

EXPANSION JOINTS AT CONCEALED COLUMN

EXPANSION JOINTS AT EXPOSED COLUMN

Courtesy of Brick Institute of America

Fig. 8.28 Expansion joints

With such a variety of structures being constructed, it is impossible to describe where every movement joint is needed. You can learn joint placement by noticing crack locations in old masonry structures, checking building codes, and using your experience. In composite walls, movement joints should be located through the wall from each other to allow both wythes to move in unison.

Common locations for movement joints:

- *At any abrupt increase or decrease in wall height.*
- *Where a wall changes thickness, such as at edges of a pilaster or a pipe chase.*
- *Above movement joints in foundations, floors, or backing walls.*
- *At every 35 feet of horizontal wall.*
- *Return angles in T-, U-, or L-shaped buildings.*
- *Joints between existing and new buildings.*

CONTROL JOINTS

Control joints for block masonry can be constructed using readily-available units, including sash and jamb blocks (see Fig. 4.7, p. 68).

Gasket material can be inserted into slots in sash blocks in a wall or at an intersection between a wall and a column or pilaster. The gasket is compressible enough to allow necessary movement and strong enough to key the adjacent elements together. The flange seals the block to prevent water entry.

Several lengths and widths of gasket are available. After laying one sash block, force the gasket into the notch. Make sure to shove the adjacent unit tight against the gasket. Lay remaining courses in the same manner, and install another piece of gasket when needed. Caulk the joint as specified when finished.

EXPANSION JOINTS

Expansion joints are constructed in brick walls by leaving a vertical joint free of mortar. If the brick is the front wythe of a two-wythe wall, the backup wythe should also have a movement joint.

Expansion joints may be backed with copper, neoprene, extruded plastic, or molded foam rubber or plastic (see Fig. 4.8, p. 68). Install the backing material while laying the wall. After laying the first course, insert the joint material into the joint from the inside. Brace the material if necessary. Lay succeeding courses until reaching the top of the joint material. Then lay the next joint section shingle-fashion (so it overlaps the lower joint material to channel water to the outside). Finally, caulk the exterior.

SEALING MOVEMENT JOINTS

Movement joints must be sealed with caulking after placement, generally after the wall is completed. Elastic sealants are designed to expand and contract as much as 50 percent of the joint thickness, so a 1/2-inch joint may expand to 3/4 inch or contract to 1/4 inch (see *Materials, Caulking*, p. 54).

Install a backup material to close off the void so you can press the caulking tightly into place, ensuring good adhesion to the sides of the joint (see Fig. 8.29). In deep voids, install filler beneath the backup material to hold the backup at an even depth. The backup and filler also prevent the caulking from adhering to material at the back of the joint. (A sealant bonded to three sides of the joint is likely to split along one edge as the wall moves.)

Use the following procedure to seal a movement joint:

1. The joint void must have the specified dimensions. Depth should never exceed the joint width. Check with the manufacturers of sealant and joints for instructions on making the seal. After installing the filler and backing, the depth should be between 1/8 inch and 1/2 inch.

2. Prepare the joint surface for the sealant. Grind or wire brush the joint to remove dirt and excess mortar. Remove ice or water by heating with a torch until the surface temperature is above 50°F.

3. Force the compressible filler (if it is used) into the void. (In some applications mortar at the back of the joint serves as filler, but this interferes with movement in the joint.)

4. Install the backup, using a roller if available to insert the material to the specified depth.

5. Install a bond-breaker tape if specified by the sealant manufacturer. This prevents the sealant from bonding to all three sides of the joint.

6. Prime the joint if required for the sealant you are using. Soak the joint but protect the facing, as primer can discolor it. Mortar

must be fully cured before priming. Use rubber gloves when handling primer and sealant.

7. Install the sealant with a caulking gun, forcing air from the joint as you proceed. Make sure the sealant adheres fully to the joint sides but not the backup.

8. Tool the joint after it has partly set with the proper size of jointer. Use water or solvent on the tool to let it slide across the caulking. Consult the manufacturer's directions about whether to wet the jointer and when to perform the tooling.

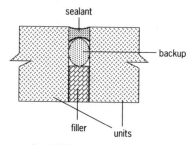

Fig. 8.29 Sealing movement joint

JOINTING

Jointing is the process of tooling mortar after the units are laid. Without tooling, tiny cracks in the mortar can allow water to penetrate and damage the joint.

The purposes of tooling are:

- *To compress and strengthen the mortar surface.*
- *To improve appearance.*
- *To create a waterproof joint.*

The following practices help insure that joints have maximum resistance to weather:

- *Material absorption, or suction, must be within reasonable limits.*
- *Units must not be moved after the bond is established.*

- *The joints should be formed under pressure so the mortar will be compressed and fully in contact with the units.*
- *The jointing must be done at the proper stage of setting—after the mortar has set somewhat but before it becomes too firm.*

TYPES OF JOINTS

Although a great variety of joints are available, the concave and V have the best water resistance for use in areas with much weather exposure. Other joints may be used indoors, where the joint is sheltered, or in other places where weathering is not a problem (see Fig. 8.30).

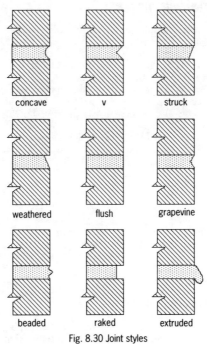

concave   v   struck

weathered   flush   grapevine

beaded   raked   extruded

Fig. 8.30 Joint styles

CONCAVE AND V

Concave or V joints offer the best protection from weathering due to their shape and the amount of compaction during jointing. These joints have the fewest cracks of any joint styles. V joints create a sharper shadow than concave joints and are used when the pattern of units is to be emphasized.

Use an S jointer to tool the head joints first, then slide a sled runner across several units at a time to make the bed joints. Compared to smaller jointers, sled runners make straighter joints, work more quickly, and put less strain on the hand.

STRUCK AND WEATHERED

Both struck and weathered joints are formed by running a trowel against the joint. For this reason they are called trowelled joints. These joints are moderately compacted during tooling and thus have some resistance to weather. Both joints are used for their shadow effect or to match existing brickwork.

A struck joint has a small shelf that allows dirt and water to gather and is subject to more weathering than a weathered joint. Weathered joints are the best trowelled joints because the face slopes to allow water to run off.

To make struck or weathered joints, slide a trowel along the joint at the proper angle when the mortar is ready for tooling. Use the entire length of the pointing or brick trowel to force the mortar into the joint, not just the tip.

FLUSH

A flush joint is made by cutting off the mortar with the brick trowel just after laying the units. This joint is not suited to exposure because the mortar is not compacted. Some compaction can be achieved by rubbing the joint with a rubber heel after a bit of setting has taken place. Flush joints are used on surfaces that will be stuccoed, plastered, or parged.

BEADED AND GRAPEVINE

Beaded joints have a depression with a bead running down its center. In the grapevine, a second hollow is centered in the wider depression between the courses. Both joints are created by striking with a special tool. Leave sufficient mortar to allow the bead to form fully. Take care to keep the joint at the center of the mortar. The beaded joint creates a sharp shadow effect on the wall.

RAKED

The raked joint should be used inside or under shelter because it allows water to gather. The joint is created with a raking tool set to the desired depth, usually 1/4 inch to 1/2 inch. To increase weather resistance, the joint can be compressed after raking by running a tuckpointing trowel along it. A clear sealant can also be applied to increase weather resistance.

EXTRUDED

The extruded joint is created by not cutting off the mortar that extrudes from the joints while units are laid. This joint has poor weather resistance and should not be used on work subject to water.

TOOLING

Joints are ready for tooling after the mortar has set for a short while after laying. If your thumb can leave a print in the mortar, the joint is ready for tooling. Avoid waiting until the mortar becomes too hard, because you will not be able to compress the mortar, and the mortar is likely to discolor.

Do not tool white mortar with steel tools, as this will leave behind steel particles that will rust and discolor. Use plastic, glass, or rubber tools instead.

Joints look best when the bed joints make continuous lines along the wall. This can be achieved by tooling the head joints first with a small jointer, then tooling the bed joints with a sled runner. After tooling, trim off burrs of mortar with the trowel. Wait a few minutes and brush the wall to remove the remaining debris. Some masons prefer to run the jointer again, to polish up the joint.

Joints in rubble stone work can be tooled by a piece of rubber hose, which can adapt to uneven surfaces. After tooling, wait until the mortar sets some more and brush the wall with a pail-cleaning brush to remove scrap mortar. Be sure the mortar is set enough or the brush will streak the mortar.

LANDSCAPING

Brick and block are used for a wide variety of landscaping purposes, including patios, walkways, planters, barbecues, screen walls, and garden walls. All materials used in landscaping should be chosen for weather resistance. Type M mortar and Grade SW brick are best. Joints, including collar joints, must be filled to prevent water from freezing and damaging structures. Pay attention to drainage requirements, both above and below the structure.

See *Barbecue and Outdoor Fireplace,* p. 166; *Paving,* p. 262; and *Walls: serpentine, retaining, garden, and screen,* p. 94.

LAYING TO THE LINE

The mason's line is the key to achieving straight, true courses. The line is attached to line blocks, nails, line pins, or story poles at the corners and stretched to achieve a tight, true guide for laying units. Units are laid with the top edge about 1/16 inch back from the line and even in height to it.

PLACING THE LINE

Lines are connected to leads or story poles and raised to the next course after a course is finished. Before raising the line, make sure the leads are set so they can hold the line without shifting.

The line must remain level throughout the work. Use a line level at the middle of the line or measure each lead carefully from a known point to be sure the line stays level. It's also a good practice to sight down the line to check that it is straight. If you see dips or notice the wind pushing the line out of alignment, install one or several trigs to correct the problem.

When masonry guides are used to hold the line, the leads need not be set first. Some masonry guides are calibrated with both standard and modular scales so the lines can be raised accurately. Otherwise, check often that the courses conform to the required spacing.

PLACING TRIGS

Trigs are small metal clips that are supported by bricks mortared into their final location at various points in the course. Trigs prevent long lines from sagging or being blown out of alignment by wind. Several trigs are required for long lines. Use this procedure to place a trig:

1. Sight along the line to determine the location and number of trigs needed. Mark these spots on the existing course or foundation.
2. Lay a unit in its final location near each mark. Use a level to plumb and align this unit, then check its joint spacing with a spacing rule.
3. Allow the unit to set briefly and place a trig, clip, or string around the line near this unit.
4. Weight the trig or clip on top of the unit with a brick or another heavy object, making sure that it is the proper distance out from the wall.
5. Check that the line is straight and level before proceeding.

USING THE LINE

A bit of judgment is required to obtain the desired gap between the line and the course. Avoid "crowding the line" by setting units too close—this will push the line slightly and result in bowed walls that must be relaid.

Pick up units in a way that will allow you to place them without disturbing the line. Make sure each unit is level from front to back and side to side. Align each face with the line at the top and the previous course at the bottom. Check that the head joints have equal width and line up with previous head joints. Make the necessary adjustments quickly, before the mortar sets. Do not disturb the line, as this will prevent other masons from working the same course accurately.

LINTELS

Lintels span an opening and transfer the weight of the structure above to bearing points adjacent to the opening. A good rule of thumb is to provide 1 inch of bearing on each jamb for each foot of opening width, with 4-inch bearing as the minimum. If a movement joint is found at the edge of the opening, you will have to terminate the lintel at the joint and provide clearance for movement.

Lintels can be built of brick, concrete, and steel. Wood is not often used because it is weak, tends to deteriorate in contact with lime, and can collapse in a fire.

BRICK

Brick lintels are best laid with running bond, although the bottom course can be rowlock. Shore up the lintel before starting—using a temporary brace of 2-inch lumber if the window or door frame will be installed later or is too weak to support the fresh lintel.

Brick lintels can be built by incorporating two 3/8-inch reinforcing rods in each of two bed joints. (see Fig. 8.31 a). Place two rods in the joint directly above the the lintel, and two more in a joint three or four courses above that. These joints must be at least 1/2 inch deep to allow mortar to bond the reinforcing correctly. Use a high-strength mortar. The bottom course can be constructed of stretchers or rowlocks as desired.

Another approach is to make a grout pocket in the wall by sawing units in the second course above the opening, as shown in Fig. 8.31b. From two to four rods may be embedded in this grout pocket.

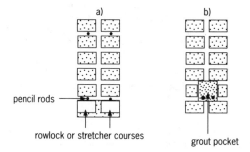

Fig. 8.31 Brick lintels

CONCRETE

Because of its strength, durability and convenience, reinforced concrete makes an excellent lintel. The end of the lintel should bear at least 7-5/8 inches on the jambs. Concrete lintels can be 1) precast or 2) built of bond-beam units on the floor or in place.

PRECAST

Concrete lintels are precast in modular dimensions for sizes that are light enough to be hoisted into place. Precast lintels are available as one-piece, full thickness units or as two-piece units, called split lintels. Split lintels are installed with a small cavity between the two pieces and are adjustable to various wall

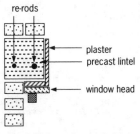

Fig. 8.32 Precast concrete lintel

thicknesses. One section is placed flush with the interior wythe, the other flush with the exterior wythe (see Fig. 8.32).

BOND-BEAM LINTEL

The special units used for lintels or bond beams can be used to cast a lintel. Use this procedure (see Fig. 8.33) to lay a bond-beam lintel in place:

1. Lay the wall to lintel height.

2. Set up the bracing. Use a 2-inch × 8-inch plank to span the opening. Choose a straight and unwarped board about 1 inch shorter than the opening. Rest this plank on dry units or 4-inch x 4-inch posts. The top of the plank should be one bed joint's height above the jamb units. Make the final adjustments for height by placing wood wedges on top of the posts. Nail diagonal braces to the shoring if needed. Nail the bracing lightly so it can be removed without disturbing the completed lintel.

3. Check that the top plank is level, square, flush with the wall exterior, and at the correct height.

4. Lay the lintel course. Use lintel or bond beam blocks for the lintel and at least 7-5/8 inches beyond the jamb. If these units have hollow bottoms, place wire mesh or building paper under end units, which do not have the plank to hold the grout.

5. Place the reinforcing steel as specified. Rods should run the entire distance of the lintel. Stirrups or other shear ties may be called for; follow directions on spacing them. Tie the rods in place.

6. Pour the grout and puddle or vibrate it into place.

7. Lay the next course of lintel blocks if a two-course lintel is required, following the above procedures.

8. Allow the lintels to cure and then carefully remove the shoring.

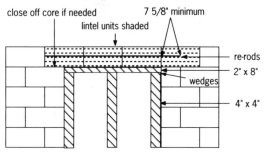

Fig. 8.33 Bond beam lintel

Bond-beam lintels can also be laid on the floor and hoisted into place. Follow this procedure:

1. Find a flat section of floor that will not be needed for a few days. Snap a chalk line across the floor to mark one side of the lintel.

2. Lay the lintel units upside-down and bond them into place with mortar.

3. Insert the proper number of re-rods into the beam and clip both ends. Tighten the clips and allow the beam to set at least twenty-four hours.

4. Insert extra reinforcement if needed and carefully turn the lintel over. Pour grout and allow to set for at least one day (longer in cold weather).

5. Hoist the lintel on the wall, position carefully, and continue laying the wall.

STEEL

Either angle iron or I-beams can be used for lintels. For a wall with two wythes, angle irons can be doubled up as shown in Fig. 4.9, p. 82. Steel lintels must bear at least 4 inches on each side. In heavier construction, this bearing is increased up to 12 inches. Because steel expands more than masonry, allow room for expansion at lintel ends.

You may need to clip the rear of the bricks that rest on the lintel to allow them to fit properly. Remove enough material so the brick can be placed easily and accurately into a good bed of mortar.

Flashing should be installed above steel lintels. Install weepholes to allow accumulated moisture to escape the wall.

Use this procedure to install an angle iron lintel over a small opening:

1. Build the wall to lintel height. Alter the bond if needed so the lintel will rest on jamb pieces longer than 4 inches.

2. Saw the angle iron (a 3-1/2-inch by 3-1/2-inch by 1/4-inch angle iron is commonly used for small openings), 8 inches longer than the opening.

3. Spread mortar on the bearing surfaces and center the angle iron on it.

4. Check that the lintel is level in both directions and parallel to the wall face before the mortar sets. The lintel may be set back 1/2 inch to 3/4 inch so it does not force the bricks to protrude from the wall.

5. Install flashing on the lintel as specified.

6. Lay units from both sides of the angle iron toward the center, making sure not to disturb the lintel. Provide weepholes if needed.

7. Shore under the lintel for at least twenty-four hours. If working in cold weather or placing heavy loads on the lintel, leave the shoring up for longer.

I-beams are used for large lintels in commercial and industrial buildings. Suspended plates may be welded to the I-beam to hold masonry units below the beam. A bearing plate may be specified at each point where the beam rests on supports.

MODULAR PLANNING

Masonry units and most other building elements are constructed to modules of 4 inches and 8 inches. The goal is to allow building components to work together without a great deal of cutting and fitting. Modular building begins with the overall size and layout of the structure. Doors and windows can be installed with a minimum of cutting, which speeds construction, and lowers costs. Because most building products are designed to conform to the modular system, tradespeople may save time and money when working on a modular building.

Some units, particularly bricks, are not always made in modular sizes. These "nonmodular" units can still fit the modular system but more planning is required.

MORTAR

Mortar is a mixture of portland cement, lime, sand, and water that is used to bond masonry materials together into structures. The cement provides strength, the lime provides workability and strength, the sand provides a framework for the cement, and the water makes the mortar flow and starts the hydration reaction by which cement gains strength (see *Materials, Mortar*, p. 68).

Masons rely on a steady stream of freshly mixed mortar of the proper grade and consistency. Any problem in the preparation or delivery of mortar will slow up the crew and/or reduce the quality of the work. Therefore, a consistent and efficient mixing setup is required.

BATCHING INGREDIENTS

A good mortar is made from an accurate mix of ingredients. During the process called batching, ingredients are measured (usually by volume) and prepared for mixing.

Mortar formulas are given as proportions of portland cement, lime, and sand. For example, a 1:1:6 mix contains one part portland cement, one part lime, and six parts of sand. When using masonry cement, which is a mixture of portland cement and lime, the usual ratio is 1:3; one part masonry cement to three parts sand.

Sand changes density as it gains or loses moisture, so it should be kept covered while it is being mixed to keep its moisture level constant. Because cement, lime, and water occupy space between the grains of sand, the total volume of a batch usually about equals the volume of sand it contains.

The standard method of measuring sand is by shovelful; however, this is not very accurate. Measuring by weight is much better. If this is not feasible, the batcher should count the number of shovels required to fill a box holding one cubic foot of sand, and multiply by whatever number of cubic feet are required in each batch.

DRY BATCHING

Dry batching is a relatively new technique for premixing mortar dry for delivery to the job site. The technique, suitable only for large projects, has these advantages:

- *There is no need to adjust the mixture in response to different levels of moisture in the sand.*

- *Mixes are more uniform.*

- *Labor and space requirements at the job site are reduced.*

HAND MIXING

Hand mixing is usually reserved for small jobs or when small batches are needed for details. Use the batching suggestions above and these procedures to hand-mix mortar:

1. Measure dry ingredients carefully into a wheelbarrow or mixing box.

2. Mix the dry ingredients thoroughly, layering the sand and cement.

3. Push the dry ingredients to the far side of the container.

4. Add mixing water to the front of the container, and gradually heap all the ingredients back toward you to form alternating layers of dry material and water.

5. Begin mixing the entire batch. Add water if needed and mix until you see no more dry spots. Continue mixing until the material is smooth.

6. Allow the batch to stand five minutes, then remix. The remixing will allow the mortar to stand longer before retempering is needed.

MACHINE MIXING

Machine mixing is used on most jobs. Although there are many preferences on how to charge the mixer, this is a common and workable procedure:

1. Start the mixer and add almost all required water. This will prevent dry materials from caking to the side.

2. Add one-third of the sand, then all the cement, then the remaining sand, then more water if needed.

3. Mix for a minimum of five minutes.

4. With the machine still running, dump the contents into a wheelbarrow, buggie, or mortar box for delivery to the masons.

Safety for power mixing

1. Keep the grate on top of the drum enclosed whenever the motor runs.

2. Use goggles and respirator while handling dry materials.

3. Do not reach into the drum unless the motor is shut off and the blades are disengaged.

4. Do not allow a shovel to reach through the grate.

CHECKING THE MIX

Thorough mixing is important. When mixing is complete, neither water nor dry material should be present in the container. Colored mortars must be mixed thoroughly to ensure color uniformity. Insufficient mixing produces weak, unworkable mortar that retains too little water and needs retempering rather quickly. However, excess mixing may reduce the strength of the batch and introduce excess air.

Check mortar by stringing out a few trowelfuls. If the mortar adheres to the trowel during handling but strings off onto the joint, it has the proper amount of cement. If it falls off too early, it has too little cement. If it sticks to the trowel instead of falling off at the proper time, it has too much cement.

A good mix must reflect working conditions. Mortar must be soft enough to work but stiff enough to hold units placed on it. Highly absorbent units may need extra mixing water. Units with little or no absorption, such as glass block and some types of stone, use a stiff, dry batch. In extremely cold weather, mortar needs far less water—perhaps only what is in the sand.

RETEMPERING

Retempering is the act of adding water to mortar that has become unworkable due to evaporation. Wooden mortar boards that absorb moisture from the mix increase the need for retempering. Retempering is needed most often on hot and dry days.

Be wary of retempering mortar that is more than two hours old because it may have hardened due to excessive hydration, not to evaporation. Compressive strength may fall as much as 25 percent by retempering a batch that is three hours old. Within about two hours of mixing, retempering will restore the bonding power of mortar even if it sacrifices some compressive strength. After 2 to 2-1/2 hours, throw out the mortar instead of retempering.

The frequency of retempering can be reduced by shading the mortar from the sun and covering it to prevent evaporation. Retarding additives are another way to reduce or eliminate the need for retempering. Some building codes and specifications contain guidelines for retempering. It is good practice to mix small batches so little retempering is needed. Colored mortar should not be retempered because it will lighten the mortar.

MOVEMENT, DIFFERENTIAL, AND CRACKING

Differential movement is the shifting of adjacent building elements in response to changes in temperature, moisture content, stress, and other factors. Movement can also result from the settling of a foundation. Because materials respond differently to these effects, the designer must plan to allow materials to move relative to each other. Otherwise, the building will develop cracks.

Building elements in direct sun get much hotter than ambient temperatures, causing a great deal of thermal movement. A 100°F increase in temperature causes a 100-foot brick wall to expand by 3/8 inch in length. Cracking as a result of differential movement can take place in these areas:

Long walls: Due to expansion from heat and moisture. Controlled by movement joints.

Foundation: A brick wall can expand at the same time that the concrete foundation shrinks from curing. This can be controlled by bond breaks between the foundation and the wall.

Parapet wall: These are a frequent source of problems because of the exposure to moisture, temperature extremes, and roof expansion.

Offsets and setbacks: Cracking can result because of differential movement.

Settlement: Can be a problem with poorly designed foundations. Settlement cracks are usually of unequal width from top to bottom.

Curling concrete slab: Slabs poured directly on masonry can curl due to shrinkage or creep. This can crack masonry that rests on the slab. Can be controlled by horizontal movement joints.

PAVING

Masonry paving is an attractive material for entryways, parks, patios, hearths, and driveways. Paving may use the pattern bonds described under *Bonds, Pattern*, p. 179. Two basic types of paving are used: mortared and mortarless. Each has advantages and disadvantages:

Paving style	Advantages	Disadvantages
Mortared	Has tolerance for different size units Attractive—harmonizes with adjacent masonry Resists water entry	Subject to cracking Harder to lay
Mortarless	Can shift slightly without showing cracks Easier to lay because no tooling and less cleanup is required.	Weaker than mortared paving. May not harmonize with adjacent masonry Cannot adjust for variations in unit size.

Exterior paving is subject to water damage, which can cause efflorescence, freezing damage, stains, and fungus or moss growth. For this reason, the paving must be pitched about 1/8 inch to 1/2 inch per foot to ensure water runoff. Driveways and streets must have a raised crown in the middle to allow drainage. During layout make sure that the drainage water has some place to go or it may puddle at the edge of the paving. Use grade SW brick, which is well-burned, for maximum resistance to weathering.

Proper slope must be considered during layout and construction. For example, if the slope is 1/8 inch per foot, then a 6-foot square will slope 6/8 inch or 3/4 inch. When paving areas in sections, such as squares, lay out the slope by setting corner units at the proper height first and using a straightedge between them to control slope. An alternative is to attach a bevel with the proper dimensions to a 4-foot level. If the slope is 1/8 inch per foot, affix a tapered piece 48 inches long and 1/2 inch thick at one end to the level. The level can be used with the bevelled side down to control the sloping dimension and the flat side down to level the dimension perpendicular to the slope.

Preparation of the base is the key to a good job. Choose a well-drained location and pitch the paving properly. In poorly drained or uncertain locations, provide gravel below the concrete (in mortared paving) or extra gravel in the base (in mortarless paving). Sandy or silty soils generally need no extra gravel for drainage.

An edging should surround the paving to prevent the surface bricks from shifting. Mortar a soldier course into place or embed it in the concrete slab. Use a straightedge and/or line to keep the paving true and level.

MORTARED

Also called rigid paving, this style should rest upon a reinforced concrete base. Mortar for brick in contact with earth should be Type M. Above-grade landscape work can be mortared with Type S cement. Lay the pavers into a 1/2-inch bed of mortar. Wide joints can be buttered as the work progresses. A slightly weaker alternative, especially useful for thin joints, is to lay the brick into the mortar bed but leave the joints empty. When all pavers are laid, mix dry sand and cement and sweep it into the joints. When the joints are full, spray the area with water until bricks and mortar are thoroughly wet. A third method is to keep the joints empty and tuckpoint after the paving is completed. Keep the area damp during hot, dry conditions for better curing (see Fig. 8.34).

a) Mortared

b) Mortarless

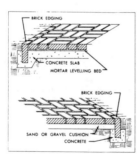
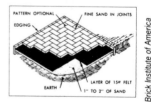

Brick Institute of America

Fig. 8.34 Mortared and mortarless paving

MORTARLESS

Mortarless paving is also called flexible or dry paving. The pavers are set on a compacted base of at least 1 inch of crushed rock or sand. If mortarless paving is laid in interiors, it can be laid on a slab or wooden framing. Asphalt roofing felt is used under the pavers.

Nominal sizes are irrelevant for mortarless construction, so be sure to study actual sizes before starting. Some manufacturers produce pavers in sizes of 4-inches to 8-inches × 1-1/2-inches to 2-1/4-inches thickness. Some of these pavers are concrete made to resemble brick, while others are fired clay.

Use this procedure to lay a mortarless pavement on sand:

1. Lay out the area and mark off the squares or other shapes for the pattern (see *Bonds, Pattern*, p. 179).

2. Spread 1 inch to 2 inches of sand over a level bed and screed it carefully. Run a plate compactor across the sand and check the level. For this reason, some masons omit the layer of felt. Continue adding sand and compacting until the base is truly level.

3. Place a layer of roofing felt over the sand to prevent weeds from growing through the base. A layer of felt will prevent the sand from working up into the gaps between the brick.

4. Set a band of edging bricks to prevent the paving from shifting.

5. Place the bricks in the desired pattern. To avoid damaging the pavers when placing them, tap them with a short length of 2 inches × 4 inches or the handle of the brick hammer.

6. Run the plate compactor across the bricks.

7. Sweep fine sand across the surface and into the joints.

Mortarless brick can also be laid over wood flooring, but this might be prohibited by building code. Joists are best placed 12 inches O.C. Lay two layers of roofing felt in opposite directions with the edges butted. Then place the pavers and sweep fine sand into the joints. Nail an edging to keep the pavers from shifting.

Mortarless paving is also laid over concrete slabs. Install a vapor barrier under the slab before pouring.

Mortarless paving can be laid over a coat of bituminous setting bed. This application is used in public areas where the bricks must bond to the base and a mortared paving assembly might crack.

PIERS

A pier is a type of masonry column that is used to support and stabilize masonry structures. All piers are columns, but not all columns are piers. Freestanding piers are used for decoration or to support floors. Piers are also used in outdoor pier and panel walls. Piers are distinguished from masonry columns because they are often thicker and are not necessarily designed to bear a load (see *Walls, garden*, p. 105; and *Columns*, p. 205).

In general, one horizontal dimension of a pier should not exceed the other dimension by more than a factor of four. The height of an unsupported solid pier should be no more than ten times the shorter horizontal dimension. Thus a solid pier 12 inches by 24 inches in cross-section could be no more than 120 inches (10 feet) high. Unsupported hollow block piers should be no taller than four times their shorter horizontal dimension.

Piers are constructed of block and brick using similar techniques. Block piers usually are constructed of pier block units, which have flush ends. Some of these units are slotted in the center for easier cutting. Brick piers are built of many types of brick. Piers can be reinforced or not, depending on the materials, application, and blueprints. If a pier is to be joined to a wall, both should be laid out and built at the same time and tied together.

BUILDING AN UNREINFORCED PIER

Use the following procedure to lay an unreinforced, hollow pier:

1. Mark off pier centerlines and exterior lines on the foundation, using a chalk line and framing square.

2. String the units dry to check bond and layout. Check that the actual dimensions match the blueprints.

3. Square the units and mark head joint locations.

4. Lay up the pier. For a small pier, lay one side or a full course at one time. Check alignment and plumb with the level. For larger brick piers, lay leads first. Block piers generally do not use leads; build entire courses or sides in one operation.

5. If you are bonding the pier to adjacent masonry with wall ties, install them as needed. If you are using a masonry bond for this

purpose, lay the pier as a lead and fill in panel sections using a mason's line.

BUILDING AN UNREINFORCED, SOLID PIER

Solid piers are used for loadbearing applications, such as supporting a lintel or beam. Procedures are similar to those for a hollow pier, except that bricks or bats are mortared into the center of the pier as the work progresses. Check the alignment of the entire course after installing the filler pieces.

BUILDING A REINFORCED PIER

Brick and block piers may be reinforced with vertical rods and horizontal reinforcing. Horizontal reinforcing may be placed in joints and/or the grout cavity. Vertical rods should extend from the foundation to the top of the pier. In some applications, rods are inserted into the footing before it is poured. Vertical rods should be secured so they remain correctly located while the grout is poured. Use standard low-lift procedures for the grouting.

Use the following steps to build a reinforced pier:

1. Mark off the wall and pier locations on the foundation or ground.

2. If the footing has not been poured, use one of two methods to secure the pier to the footing: a) Cut the vertical reinforcement about 6 inches to 8 inches taller than the pier. Insert the rods in the ground so they will terminate just below the pier cap, and brace the rods in position. b) Insert pins into the footing that will protrude far enough into the grout. Then pour the footing.

3. If the rods protrude only a few inches from the footing, cut vertical rods a few inches shorter than the pier and brace them into position.

4. Lay the first course of the wall and pier.

5. Place horizontal reinforcement on the first course or in the cavity. If 1/4-inch (pencil) rod is placed on the course, make the bed joint 1/2 inch deep to allow proper bonding of the rod.

6. Continue laying courses, using proper bond and placing reinforcement at desired courses.

7. When a few feet of the pier is finished, allow it to set until firm. Then puddle grout into the cavity with low-lift grouting procedures.

8. Continue building to the full height, grout again, and install the cap.

INSTALLING BEARING PLATES AND ANCHORS

Loadbearing piers can be topped with steel bearing plates and/or anchor bolts. A bearing plate is designed to enlarge the cross-section of the beam to spread the load around. The plate is centered on the pier, mortared into proper position and height, and levelled. Beams and lintels can be welded to the plate for stability. Consult specifications.

Steel templates are steel plates (often 1/4 inch thick), which can be inserted into joints to increase the bearing strength of a pier. Cut the steel so it is at least 1/2 inch smaller in both horizontal dimensions than the pier. When you have reached the proper course height, lay a 1/4-inch bed of mortar and place the plate. Level the plate and lay another 1/4-inch bed of mortar on top of it. Lay the succeeding courses. Because this bed joint will be close to 3/4 inch thick, you may have to adjust the course spacing to make the pier conform to modular spacing.

If required, insert angle bolts into the grout to the proper depth before placing the bearing plate. In ungrouted piers, insert the bolt into the collar joint or a grout pocket supported on metal lath. Then mortar the bearing plate into place and slip it on the bolts. Do not tighten the bolts until the mortar has set.

PILASTERS

Pilasters are sections of wall designed to support loads, increase lateral stability, and terminate walls next to an opening. Many pilaster designs are possible. Reinforced pilasters can be as narrow as the wall itself; unreinforced pilasters are thicker and must project from one or both wall faces. The details are described on blueprints and specifications.

Because they are likely spots for differential wall movement, pilasters commonly incorporate movement joints. Special blocks are used for this application.

LAYING A MASONRY BONDED PILASTER

A masonry bonded pilaster must have at least one unit per course bonded to adjacent walls. The pilaster is usually laid at the same time as the walls. Use the following procedure:

1. Lay out the wall lines and mark the pilaster location on the foundation. Square the lines with a square.

2. String the units dry for the pilaster and adjacent wall. Make sure the pilaster will bond properly with the wall in one or both alternate courses.

3. When the spacing is correct, mark the head joint locations on the foundation.

4. Cut any units needed.

5. Place and align the first course. Start with the exterior wythe. Then lay the interior wythe and any bats required for the pilaster. When the pilaster course is finished, ensure that it is level with the wall by stretching a mason's level from the pilaster to the wythe.

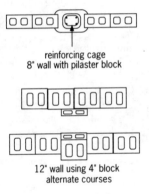

reinforcing cage
8" wall with pilaster block

12" wall using 4" block
alternate courses

Fig. 8.35 Pilaster designs

6. Install joint reinforcing in pilasters that will be grouted to help the pilaster resist the hydrostatic pressure of the fresh grout.

7. Use the story pole to control course height.

8. Place a bearing plate and/or angle bolts as described on p. 268.

LAYING A REINFORCED PILASTER

Reinforced pilasters may be constructed of block or brick. Sometimes the pilaster alone is reinforced, other times the wall is also reinforced. Consult detail drawings for each application. A reinforcing cage can be used to provide both vertical and horizontal reinforcement within the grout cavity of a block pilaster. This cage is generally installed before the pilaster is constructed. If a movement joint is used, make sure not to extend horizontal reinforcement through the joint unless specified.

Reinforced block pilasters often employ different block types in the alternate courses as shown in Fig. 8.35. Pilasters can also be constructed from two-core units with flush ends. (Three-core units can be used but it is more difficult to line up the cores to insert rods and grout.) A reinforcing cage is not used with these applications, but joint reinforcing can be.

Use this procedure to lay a reinforced pilaster, following the above procedure through step 4:

5. Check that reinforcing will fit in the cavity without touching the inside of pilaster blocks.

6. Place and align the first course.

7. Insert the reinforcing cage if one is used and brace it into position. Lay horizontal reinforcing in the proper courses.

8. Lay up leads at the wall corners, using the story pole to control course height. Then lay the pilaster with the wall courses.

9. When each pilaster course is finished, ensure that it is level with the wall by stretching the level between the pilaster and the wythe.

LAYING A PILASTER WITH A MOVEMENT JOINT

Pilasters can be connected to a wall by a movement joint. Construction is similar to the reinforced pilaster (above) except no reinforcement should bridge the movement joint. If reinforcing wire does bridge the joint, grease the wire before installation to prevent it from bonding to the mortar. Rake the joint at least 3/4 inch deep after completing the pilaster. Then caulk as usual for a movement joint.

An expansion joint may be formed next to a brick pilaster by allowing the wall bricks to penetrate into a chase in the pilaster. The bond must be broken so movement can take place. Notice that a compressible filler is used at the termination of the wall bricks (see Fig. 8.36).

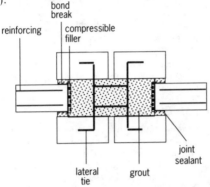

Fig. 8.36 Brick pilaster with expansion joint

REINFORCING

Reinforcing is the use of metal elements to strengthen masonry structures. Masonry walls employ reinforcing steel to increase resistance to stress. The use of reinforcing allows lateral supports, such as pilasters or other building elements, to be placed at greater intervals, thus increasing design flexibility. Reinforcing also allows masonry buildings to meet building codes in areas prone to earthquakes. Reinforced construction may be applied to walls, chimneys, pilasters, columns, and piers.

Three types of building element can be used singly or in combination to create reinforced masonry construction: 1) bond-beams; 2) grouted cores or collar joints; 3) joint reinforcement. It is the architect's responsibility to specify the type and placement of reinforcing. However, masons should be familiar with certain design and installation criteria:

1. Cavities should be large enough to allow steel to be completely embedded in grout or mortar.

2. Reinforcing steel should be surrounded with grout or mortar.

3. Grout pockets and bond beams should be puddled to eliminate voids.

4. Steel must be held in position during pouring.

5. Re-rods or joint reinforcement must overlap the correct distance at splices.

Other information pertaining to reinforcement is found under *Grouting*, p. 240; *Concrete block construction*, p. 206; *Columns*, p. 205; *Piers*, p. 266; *Pilasters*, p. 268; *Chimneys*, p. 128.

JOINT REINFORCEMENT

Steel pieces of various shapes can be used to strengthen masonry bed joints. Joint reinforcement can also tie two wythes of a wall together, thus replacing wall ties and anchors. When sections of joint reinforcement meet, the lateral wires must overlap at least 6 inches to ensure adequate tensile strength.

Prefabricated corners and tees may be preferable to job-fabricated pieces because they save labor and are more accurate and stronger.

PLACEMENT

Placement of reinforcement is specified by blueprints. However, these guidelines are helpful for builder-designed structures:

- *Joint reinforcement should not extend through a movement joint.*

- *Provide reinforcement in the first and second bed joints below windows, and above windows and doors. This reinforcement should extend 2 feet past the opening in both directions, unless a movement joint is closer than this.*

- *Provide reinforcement every 16 inches in wall height (except within 24 inches of a bond beam).*

REPAIRS

The basic procedure in repairing masonry is to remove the decayed or damaged area, clean off the surrounding material to allow a good bond, and repair with a material that matches the original. When making a repair, take action to remove the source of the damage if possible. Thus, if spalling resulted from faulty flashings, repair or replace the flashing to prevent future water damage. If the repair was required because the building shifted due to frost heaving, make the proper improvements to the grade or drainage system to correct the underlying problem (see *Coloring Mortar*, p.204; *Troubleshooting*, p. 293).

REPLACING BRICKS OR BLOCKS

Remove all damaged units and decayed mortar, taking care not to disturb any sound structures. Use a saw if necessary to make a clean cut. Clean the opening, dampen it to reduce absorption if the units have high suction, and mortar the opening on all sides. Use preshrunk (prehydrated) mortar (see below). Shove the unit into place. You can be sure you have used enough mortar if it squeezes from the joint.

TUCKPOINTING

Tuckpointing is the repair of eroded or damaged joints without replacing the units. This damage generally occurs because water invaded the joint and subsequent freezing weakened the mortar, although it can also result from mortar that was poor in the first place or from the shifting of a building.

First clean the joint thoroughly with a hammer and chisel. The size of chisel depends on the size of crack. Remove mortar at least 1/2 inch to 3/4 inch deep. If the decay extends deeper, remove that mortar also. Tuckpointing rakers are also used for mortar removal. On large jobs, a tuckpointing grinder can be used. Grinders usually take a toll on the edges of the units and are not suggested for preservation work.

After chiseling or prying out rotten mortar, blow out the waste with compressed air, wash it out with water, or remove it with a stiff brush. Wear eye protection while grinding, chiseling, and removing mortar.

Wet the joint after cleaning (unless water was used in washing) to prevent the surrounding masonry from absorbing too much water from the mortar. Use a mortar best suited to the brick being repaired (see *Coloring Mortar* for information on coloring mortar to match the surrounding masonry, p. 204).

Prehydrate the mortar to reduce shrinkage in the joint by mixing sand and masonry cement dry in the mixing box. Add just enough water to make a doughlike ball and mix. Wait one or two hours and add more water to make the mortar workable. Tuckpointing mortar should be drier than normal mortar and should shrink less while setting.

Place the mortar on a hawk or the bottom of a brick trowel and force it into the joint with a tuckpointing trowel. Fill deep joints gradually, giving the mortar time to set to prevent excessive sagging or cracking. When the joint is full, tool it to match the surrounding wall.

SITE PREPARATION AND LAYOUT

Site preparation is the process of making a site ready for builders to work. Site layout is the act of determining exactly where a building will be placed on the site.

The allocation of site preparation and layout tasks varies with the job. A crew from the architect, surveyor, carpenter, or general, concrete, or masonry contractor may prepare a site. No matter who does the preparation and layout work, the masons should ensure that it was done correctly before starting their work. When working in or near an excavation, make sure to observe safety precautions for trenches and excavations.

The first drawing to examine is the site plan, which gives the building height and location in relation to specific reference points. Other necessary information may be found on the blueprints and specifications.

The following checklist for site preparation includes tasks that might need to be done before building starts:

    Preliminary work
    Estimate
    Contract
    Check credit or bank of customer, general contractor, and
        subcontractor
    Survey

Percolation test
Building permit and zoning approval
Check with city engineer or agriculture department about
    possible soil, subsoil, or groundwater problems
Engineering tests
Site access and task allocation
Accessibility from road—temporary road needed?
Temporary water—is hauling needed?
Temporary electricity—is generator or additional wiring
    needed?
Location for storing materials and equipment
Distance to job site
Distance for material delivery—Any cartage charge?
Distance to dump site—who is responsible for hauling refuse?
Safety and sanitary requirements of the job
Staking out
Temporary covering of site
Excavation—who is responsible?
Welding equipment available if needed?
Sequence of construction understood?
Additional drawings needed?

LAYOUT

Laying out the building is a key task of site preparation. Layout
requires finding the height and location for batter boards,
excavation, footings, and foundation walls. Lay out a site with
extreme care or else the building might be incorrectly located.

"Shooting heights" is the process of figuring out the relative
elevation of various locations on a building site, and is generally
before and after excavation. Use either a builders' level or a transit
to shoot heights for excavation and locating footings. Shooting
heights is relatively easy on flat terrain, but more complicated on
hillsides.

Layout work begins by referring to a "reference point" or
"reference stake," which is established by the architect or engineer.
The reference point can be a stake, a mark in the road, a hydrant, or
a building corner. A reference stake may be measured from the lot
line or a road or set by an architect or surveyor. "Target" stakes are
located at spots whose height you want to find.

From the reference point you can find the location and height of one building corner with either of two methods: 1) With the level set up directly over the reference stake, sight the target stake and measure the difference in height. 2) Set the level up between the reference stake and the target, and transfer the height from the reference to target.

SHOOTING HEIGHTS WITH A BUILDERS' LEVEL OR TRANSIT

Use the following procedure to shoot heights from a position between the reference point and the target. During this procedure, it is easiest to assume that the reference stake is 100 feet high. This will prevent a need to assign negative values to elevations below the reference point:

1. Mount the level on the tripod at a position approximately halfway between the reference point and the target. Roughly level the builders' level while setting up the tripod. Make sure the tripod legs have a firm purchase on the ground, especially on paved or muddy surfaces.

2. Center one of the two bubble levels by adjusting one pair of opposing screws on the plate of the tool. Repeat with the other pair of plate screws. Check that the first pair is still level and continue until both bubbles are exactly centered.

3. Have an assistant place the bottom of the leveling rod on the reference point and hold the rod vertical. Sight on the leveling rod with the telescope. Have the assistant slide the target up or down the rod until it is directly in the telescope's cross hairs. Check that the level is still perfectly horizontal and the rod is vertical. Tighten the clamp screw on the telescope.

4. Record this reading (5.5 feet in Fig. 8.37) and add it to the assumed height of the reference point, 100 feet. The height of the telescope is 105.5 feet.

5. Have the assistant move the leveling rod to the target position. Loosen the clamp screw, pivot the level, and sight on the leveling rod. Read the sighting and record it (this equals 4.3 feet in the example). This shows the telescope is 4.3 feet above the target stake.

6. Calculate the relative height of the target. Subtract the height

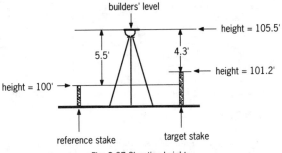

Fig. 8.37 Shooting heights

from the telescope above the target from the height of the telescope. 105.5 − 4.3 = 101.2.

7. The height of the target is said to be 101.2 feet, although it is actually 1.2 feet above the reference point. If the second sighting was 7.8 feet, the calculation would be as follows: 105.5 feet − 7.8 feet = 97.7 feet. This is why it is handy to assume that the reference stake is at an elevation of 100 feet.

8. For shooting heights on uneven terrain, several sightings must be made. Refer each successive sighting back to the reference point as you work.

SHOOTING ANGLES WITH A BUILDERS' LEVEL OR TRANSIT

Builders' levels and transit are also useful for finding angles between two points when laying out buildings. Use this procedure (see Fig. 8.38) to find an angle:

1. Set up the level or transit directly above the reference point or a corner stake. Mount the plumb bob and move the instrument until the bob rests directly above the reference stake.

2. Level the instrument as described above.

3. Loosen the clamp screw and sight on a reference point.

4. If the tool has a movable horizontal scale, loosen the clamp and set the scale to 0°. If not, record the exact reading on the horizontal scale.

5. Loosen the clamp screw and pivot the telescope to the target. If the movable horizontal scale was set to 0° in step 4, simply read the new angle on the scale. If not, subtract the second reading from the first reading. Most layout will start out by finding right (90°) angles. If the tool has a vernier scale (see below) use it to measure an angle more exactly.

A vernier scale is a device that measures extremely accurate angles and lengths. Vernier scales on builders levels and transits have 12 divisions on each side of the 0 indicator. This allows an accuracy of one-half of a degree, or 5 minutes. Use this procedure to read a vernier scale:

1. Read the degrees opposite the 0 indicator on the vernier scale. If the 0 is perfectly lined up with a degree mark, then the reading is exactly that many degrees. Otherwise you must also read the vernier scale.

2. Round down from the 0 indicator to find the degrees scale.

3. Find the vernier mark which lines up best with *any* mark on the degree scale. Note the minutes on the calibration next to that vernier mark.

4. Add the minutes on the vernier to the degrees previously found to get the exact reading.

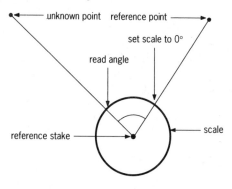

(plan view)

Fig. 8.38 Finding angles with the builders' level or transit

FINDING THE LENGTH OF THE DIAGONAL

The length of the sides of any right triangle (a triangle with one 90° angle) conform to a relationship called the Pythagorean theorem. In right triangle ABC, the relationship between the length of the sides is $A^2 + B^2 = C^2$. In other words, the length of side A multiplied by itself plus the length of side B multiplied by itself equals the length of side C multiplied by itself. This is always true if C is the hypotenuse, the side opposite the right angle.

To find the length of the diagonal (C) of a 32-foot × 40-foot building (ABCD):

1. Multiply the length of one side by itself: $32 \times 32 = 1024$

2. Repeat for the second side: $40 \times 40 = 1600$

3. Add the products: $1024 + 1600 = 2624$

4. Using a hand calculator, find the square root of the sum. R(2624) = 51.22 feet. This is the length of the diagonal.

THE 6–8–10 METHOD

Before starting construction, check that all right angles are truly square by checking that diagonals are equal or using the 6–8–10 method:

1. Measure 8 feet out from one square corner and mark.

2. Measure 6 feet in another direction from that corner, and again mark.

3. Measure the hypotenuse, the distance between the two marks. If it measures 10 feet, the angle is 90°.

USING A TRANSIT

A transit can do everything a builder's level can and more because its telescope can swing away from horizontal. This allows use of a transit to determine whether building lines are vertical and to lay out stakes in perfect alignment with a reference point. With a transit, you can measure heights and then drop the telescope to mark exact stake locations on the ground without needing to change positions.

Use the following procedure (see Fig. 8.39) with a transit to lay out

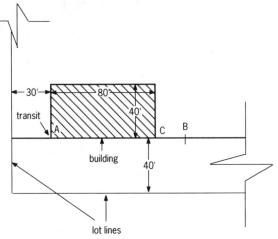

Fig. 8.39 Transit layout

a 40-foot × 80-foot building 30 feet from one lot line and 40 feet from the other lot line:

1. Following directions for the builders' level, above, set up the transit over building corner A (established by the architect or surveyor). If the building corner is not known, see *Marking building corners without a transit*, below.

2. Locate stake (B) by measuring from a known point, as specified by the architect or surveyor. In this case, (B) is 40 feet from the lot line. Make sure to locate (B) so it will be outside the building. Line (AB) is a reference line for the layout.

3. Measure 80 feet from (A) to the building corner, point (C). Find (C) exactly by lowering the telescope vertically from point (B) until you sight the end of the tape in the crosshairs. Drive a stake at this point.

4. Sight the stake with the telescope and direct your assistant to mark the exact location of the corner on it. A nail or tack is sufficient.

5. Repeat this procedure with the third and fourth corners, using the angle scale to set the proper angle between corners (or you can find the fourth corner by measuring the diagonals using the technique above).

MARKING BUILDING CORNERS WITHOUT A TRANSIT

Example: Lay out a 32-foot × 40-foot foundation on a building site 110 feet × 100 feet. The building will be set back 35 feet from the south lot line and 24 feet from the west lot line. Before starting, check the site plan and be sure you are using the proper lot lines and reference points. Check that the setbacks meet the local building codes. Then follow this procedure (see Fig. 8.40) to set the building corner stakes, also called hubs:

1. Mark the setback from the south lot line on the east and west lot lines. Measure 35 feet from (AB) and insert stakes (E) and (F). Stretch a string to mark line (EF), the south wall reference line.

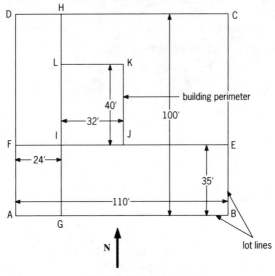

Fig. 8.40 Site layout

2. Measure the 24-foot setback from the west lot line and insert stakes (G) and (H). Then stretch line (GH), the west wall reference line.

3. Drive a stake at (I), the intersection of (EF) and (GH). This is the first hub.

4. Measure 32 feet from (I) along line (EF) to establish hub (J), the second hub.

5. Using the procedure above, find the length of the diagonals. Then stretch a tape measure the length of the diagonal (51.22 feet) from (I) toward corner (K). Stretch another tape the length of the building (40 feet) from (J) toward (K).

6. Set a stake where the two tapes meet to mark hub (K).

7. Repeat to find hub (L). Now the corner hubs are established.

8. Measure (KL) and confirm that it is 32 feet.

9. Measure both diagonals and check that they are the same; this assures that the rectangle has square corners.

ESTABLISHING BATTER BOARDS

Batter boards are temporary devices used to hold building lines during excavation and footing work. Batter boards are set after the building corners are established to allow excavation to proceed. The setback of the board depends on the depth of excavation—with a minimum of 3 feet and occasionally much more. The exact location of the batter boards is not important as long as they are level and the strings accurately attached. Although triangle-shaped batter boards are shown, each leg of the batter board can be a separate structure, which is useful when the batter board must be set back quite a way from the hub.

Use the following procedure (see Figs. 8.41 and 8.42) to locate the batter boards and batter board lines for a 32-foot × 40-foot building.

1. Decide how far back the batter boards must be from the hubs (use 3 feet for this illustration). Insert stakes (B) and (C) about 3 feet back from the foundation lines (DE) and (DF). Insert stake (A) to form a right angle with stakes (B) and (C).

2. Mark a known height on the stakes by one of these methods: a) Transfer a known height (see p. 277) from a reference point to

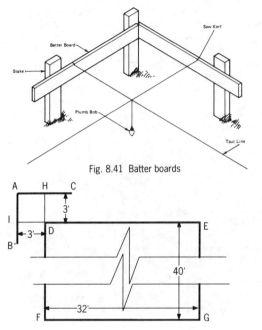

Fig. 8.41 Batter boards

Fig. 8.42 Setting batter boards

all three stakes and mark. b) Shoot the height on one stake and use a level to mark the other two stakes. c) Use a water level to transfer the height from a reference point.

3. Set each batter board so its top edge is level with the marks and nail it into position. The batter boards are now a known height and the footing can be located according to them.

4. Set batter boards at the other corners using the same procedure.

5. Pull lines (FH) and (EI). The point of intersection should directly above a hub. Use a plumb bob to transfer down to the hub. Move lines (EI) and (FH) until they are in the correct location above a hub.

6. Fasten remaining batter lines on the boards using the same procedure, and check that they intersect above the hub. When done, cut a saw kerf or hammer in a nail where each line attaches to a batter board.

7. Measure wall lengths to check that they are correct. Then recheck the angles between hubs with a steel square, builders' level, or transit. In a rectangular building you can test the corner angles by checking that the diagonals are equal.

8. Take the lines down until the excavation is finished, then replace them to reestablish hubs for setting the footings.

STEPS

Masonry steps are commonly used for garden walks and building entries. Each element of a set of steps should be described by blueprints, but check building codes before proceeding. Steps can be constructed of concrete, block, or brick. Brick and block can be used over a concrete or block foundation.

For exterior steps, treads at least 12 inches deep are preferred because they are safer, especially in wet or icy weather. For interior steps, a tread between 9 inches and 11 inches deep is adequate. A rule of thumb is that the sum of the two risers and one tread for an interior step should range between 24 inches and 25 inches.

TERMINOLOGY

Total rise: The vertical distance from the lower walkway to the upper walkway: Riser height × number of risers = total rise.

Line of flight: A line paralleling the angle of the steps. Find the line by laying a straightedge across the nose of the treads. A line of flight 10° to 15° from horizontal makes a very gentle set of steps; 30° to 35° is common. Angles steeper than that may be too steep.

Riser: The vertical part of a step.

Tread: The horizontal part of a step.

Width: The overall horizontal measurement of a step assembly; 3

feet is a comfortable width, although narrower steps may be acceptable.

Clearance: The vertical distance between a tread and any structure above it. The accepted standard is 7 feet 2 inches clearance, to allow tall people and people carrying objects to use the stairs easily.

PROCEDURE FOR BRICK STEPS

Use the following procedure to lay brick steps over a concrete foundation. Where frost is a concern, foundations should extend below the frost line. Steps poured on grade (with no footing beneath frost line) should not sit beneath a threshold because the frost can raise the steps and break the threshold.

PLANNING AND LAYOUT

If the steps adjoin a walkway paved with the same material, the top tread in the foundation must be flush with the base walkway material. If the steps are adjacent to a walkway without paving, depress the foundation tread below the walkway surface by the height of one step paver.

Notice to which side the door opens and make the platform large enough that a person can stand on it while the door opens.

The following procedure (see Fig. 8.43) describes laying out the steps, forming and pouring the foundation, and setting the units on the foundation. If the steps are described in the blueprints, proceed to step 4:

1. Determine the total rise and run. Drive stake 1 into the ground at the top of the slope and mark the top ground level on it as point (A). Drive stake 2 vertically in the ground at the bottom of the slope and check that it is plumb. Use a mason's level or line level to transfer ground level from point (A) to point (B) on stake 2. The length from point (A) to point (B) is the total run. Mark point (C) at the lower ground level on stake 2. Measure from (B) to (C) to find total rise.

2. Determine the number of treads. Divide the total run by desired tread depth. If the total run is 75 inches and you want the treads 12 inches deep, $75 \div 12 = 6.25$ treads. Since you cannot have a fractional tread, you have two choices:

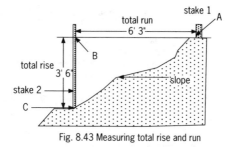

Fig. 8.43 Measuring total rise and run

a. Change the total run (and the line of flight). Do this by multiplying the desired number of treads by the desired depth. $12 \times 6 = 72$, so 6 treads 12 inches deep gives a total run of 72 inches.

b. Increase the total run and retain the line of flight. To find how deep the treads must be for 6 treads with a 75-inch run, calculate 75 inches ÷ 6 treads = 12.5 inches per tread. Choose the best alteration to the plans by checking how it will affect the final step layout. (For this illustration, the second option was chosen.)

3. Determine the rise per step by dividing the total rise by the number of treads plus one (steps have one more riser than tread). Assume the total rise was 42 inches and remember that seven risers are needed for six treads: 42 inches ÷ 7 = 6 inches.

4. Make a pitch board. See Fig. 8.44. (The steps can also be marked with a steel square—use it in the same manner as the pitch board.) Lay a carpenter's square on a scrap of plywood with the corner flush to an edge. Measure from the corner, point (A), the 12.5-inch tread depth and mark (B). Then measure from (A) the 6-inch rise and mark (C). Mark lines (AB), (BC), and (AC) and remove the square. Cut all lines accurately.

5. Select two pieces of straight 2-inch lumber for the two side forms. Each side form must be longer than the line of flight.

6. Add 1 inch + rise + slab thickness to find the height of the side form. Slab thickness is usually about half the riser

height—never less than 4 inches. (In this illustration, slab thickness is measured along a line that will be vertical when the form is positioned. Slab thickness may also be figured perpendicular to the line of flight.)

7. Mark points (A) and (B) and snap a chalk line to mark line (AB), the bottom of the slab, on the form.

8. Carefully mark the steps with the pitch board as shown. Allow some clearance so the steps do not crowd either end of the side form.

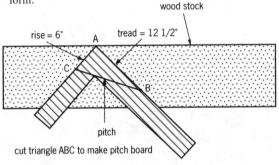

cut triangle ABC to make pitch board

Fig 8.44 Pitch board

9. Mark the top end of the form. Mark line (CD) parallel to riser (E) and one tread depth away from (E).

10. Mark the bottom of the form. Mark line (FG) perpendicular to the bottom tread (FH). Mark line (GI) parallel to (FH) to intersect (AB).

11. Stack both side form pieces and cut exactly along the lines marked. Do not allow cuts to go past the lines at the bottom of the risers; this would weaken the forms.

12. Select one form board for each riser and rip it to the exact riser height. The length of this board equals the finished step width plus twice the side form thickness.

13. Nail riser forms directly to the front of the side forms. Check width, treads, risers, total rise, and total run. Check that the form is square and nail a diagonal brace to keep it square. The form is complete (see Fig. 8.45).

LAYING STEPS

Use this procedure to install the form, pour the footing, and lay the steps:

1. Excavate the area beneath the steps. Set stakes at the future locations of the noses of the top and bottom treads. Connect a line between the stakes to mark the line of flight. Measure from the bottom of the side form to the nose of a tread to find how deep you must excavate beneath these strings. If this

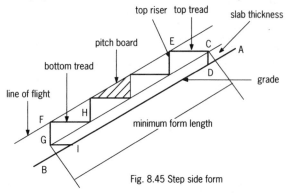

Fig. 8.45 Step side form

measurement was taken perpendicular to the line of flight, make sure to measure the excavation perpendicular. If the measurement was taken plumb, measure the excavation plumb.

2. Perform this excavation and compact the soil beneath the steps. Place reinforcement, if it will be used, according to specifications and blueprints.

3. Place the form in position. Check that the top and bottom steps align properly with the walk or building entry. (If the steps adjoin masonry walkways, the form top and bottom should be flush with the base material at the top and bottom walkways. If the step adjoins an unpaved area, set the form one thickness of paving material below grade at top and bottom. This will give the steps the proper level after the paving is completed.)

4. Level the treads and plumb the risers. Check again that the form is square. Make sure reinforcement is still in the proper location.

5. Anchor the form into position. Use enough anchors to prevent concrete from spreading the form. Coat the insides with form oil to simplify removal.

6. Pour concrete into the bottom step. Puddle it well to remove voids. Strike off the step across the riser and side forms. Continue working upwards, one step at a time, making sure that successive pours do not bulge the lower steps. Leave a rough finish on the treads to promote mortar bonding when laying the pavers.

7. Cover the foundation with tarps or straw and cure it for a few days. Fog with water in dry conditions.

8. Select bricks of Grade SW (severe weathering) for the surface. Mortar can be Type M or S, depending on conditions. On wide steps, it is best to work from the top down, so you do not disturb fresh work by kneeling on it.

9. Begin laying the top treads in the order shown in Fig. 8.46. Level the treads from side to side, but provide about 1/4 inch per foot slope toward the bottom of the steps for water runoff. Make the risers plumb.

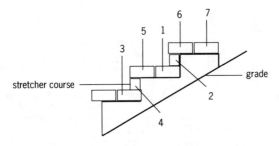

Fig. 8.46 Lay step units in this order

10. As you proceed, "rack" the noses of the steps by placing a 4-foot level or a story pole against the noses of the steps. Make corresponding joints in adjacent steps line up for appearance sake. Tool the joints well to prevent water penetration. Use a sealer if specified.

STONE CONSTRUCTION

Stone used for construction varies from sawn, rectangular units to rough field stones. Stone availability and price vary by location, although stone can be shipped long distances if needed.

While some elements of stonework resemble brick and block masonry, there are significant differences. Stone is harder to cut, and usually heavier and slower to work. The result is that stonework is more expensive than brick or block masonry. In addition, masons must pay more attention to the aesthetic impact of the wall. And because some types of stone are not absorbent, the mortar may set slowly and stones may shift during construction.

Heavy stones require spacers to prevent them from forcing mortar out of the joint. Wet wooden wedges or pieces of plastic or lead are handy for this purpose.

LAYING A VENEER WALL

The procedures for working with stone vary as much as the material itself. However, the following technique for laying a stone wall on an existing foundation can be adapted to most types of stone:

1. Spread out the stones near the work site and look them over. Think about how you want the wall to appear. Determine the thickness of the wall—usually between 4 inches and 8 inches. Roll the stones to find the best faces, and leave those faces up so you can easily locate the right stone.

2. Lay out a few of the larger stones for the base course. Observe how they contribute to the overall appearance of the wall. Try for a balanced, interesting appearance. Do not concentrate one type of stone at any location. Make sure stones protrude equally from the wall. Mark stones with chalk or heavy pencil for cutting.

3. Remove marked stones from the wall and cut them. Use a mash hammer and stone chisel and wear goggles. Begin the cut slowly and increase the strength of the hammer strokes as you work. Use a sledgehammer for larger stones.

4. Replace trimmed stones and check the fit. When the fit is correct, remove all stones and clean them by brushing thoroughly.

5. Lay a thick bed of mortar and replace the stones in their previous locations. Try for a joint of 1/2 inch to 1 inch for rough work and 3/8 inch to 3/4 inch for ashlar work. Use wedges to support heavy stones while the mortar sets. Remove the wedges and point up the gap while the mortar is still relatively fresh.

6. Point up low spots in the wall with mortar after the stones are positioned. Throw mortar in with a brick trowel or use a pointing, margin, or caulking trowel to fill the gaps. Use stone chips to help fill larger voids.

7. Dress the joints with a pointing or margin trowel. Strike them smooth with whatever tool has the best shape. A regular jointer works well for ashlar stone but a rubber hose may be needed for rubble work. Joints in stone are often raked. Use a long-bristled brush to remove scrap mortar from the work after the mortar has begun to set.

8. Continue selecting stones for the new work. Use a line or level to ensure that the wall remains plumb. If there is a backup wall, measure from it to check wall thickness. Stand back to check the appearance from time to time. Make sure to distribute the various sizes, shapes, and colors of stone around the wall. Work in several locations so stones have time to set before you must place new ones on top. Use special care when placing heavy stones and ask for help lifting them.

9. Use noncorrosive wall ties as specified. As a rule of thumb, provide one tie per 2 to 4 square feet.

10. Clean the wall when finished. Use care—acid treatments may discolor some stone, including limestone. Select the proper cleaning compound for the stone you are cleaning.

APPLYING LIGHTWEIGHT ARTIFICIAL STONE

Building codes should be consulted before starting to lay artificial stone because they may specify particular attachment devices. Use the following method:

1. Prepare the surface according to these specifications:
   a. Masonry materials: Clean masonry, stucco, and concrete need no preparation. Old or dirty masonry must be cleaned. Remove form release agents by wire brushing or washing with muriatic acid. Sandblast dirty or painted masonry. An alternative for unsound or dirty substrates is to securely fasten metal lath over the wall.
   b. Plywood or other sheathing: Cover with a weather-resistant barrier such as roofing paper, with joints lapped. Cover this barrier with metal lath secured by galvanized nails or staples 6 inches O.C. vertically and 16 inches O.C. horizontally. Keep the lath 1 inch back from exposed edges.
   c. Open studs. Apply a weather-resistant barrier, nail up metal lath and apply a 1/2-inch to 3/4-inch scratch coat. Allow to set for 48 hours.
   d. Metal buildings: Use the technique for open studs. Fasten lath with self-tapping screws 6 inches O.C. vertically and 16 inches O.C. horizontally.
2. Mix mortar to proper color and apply a 1/2-inch to 3/4-inch coat over 5 to 10 square feet.
3. Start setting stones at a corner, alternating corner pieces so the long legs extend first to one side and then to the other. Keep the mortar joints consistent.
4. Work from the top down or bottom up. Working from the top keeps mortar droppings off the lower courses. You can either press a stone firmly in the mortar bed or plaster the back of a stone before applying it to the backing.
5. Cut and trim stones if needed using wide-mouth nippers or a hatchet. When broken stones are used, cover the break with mortar or place it in an inconspicuous location.
6. Keep stone surfaces clean. Remove droppings with a whisk

broom after they become crumbly, not with a wire brush or wet brush.

When working in hot weather, moisten the back of each stone to control suction. For exterior installations, allow 4-inch clearance between grade and stone. Use flashings, copings, and other standard building practices to prevent water from penetrating from above.

LIFTING STONE SLABS

Several types of lifting devices are used to hoist large panels of stone into place. "Gin" poles, or pole derricks, are mounted on a forklift to hold a block and tackle for the stone.

C-clamps can be used to hoist slabs. The stone must have a small hole drilled into its back for the clamp to attach. Multiple clamps provide more lifting power and safety. Use pressure pads for a good grip on rough stones.

Lifting rigs slip around a notch cut into the rear of the stone. The notch must be far enough from the edge to prevent breakage.

When you lower a stone hoisted by a sling wrapped around its bottom, set it down on spacers so the sling can be removed.

Safety for lifting stone:

• *Do not attempt to hoist slabs of stone from the horizontal to vertical position with just an attachment at the top. Use slings in addition for this operation.*

• *Slings must be long enough to prevent damage to stone edges.*

• *After assembling a lifting rig, inspect it carefully before use.*

• *Perform all lifting operations slowly and safely.*

• *Use hands, wire, or rope to control the stone as it is lifted.*

• *Consult manufacturers of lifting devices for further safety procedures.*

TROUBLESHOOTING

Although masonry generally makes durable, long-lasting structures, some troubles can occur. One goal of good construction techniques is to prevent trouble. Certain repairs are possible in existing buildings, but others are not. (For information on troubleshooting a fireplace, see p. 151.)

CRACKING

Cracking often results from a settling in the foundation and is difficult to remedy. Cracks can be repaired with the techniques described under tuckpointing p. 273.

Once cracks have occurred, they can be repaired, although this is likely to be less effective than proper design and construction practices in the first place. If the crack is straight, saw it out, install a control joint, and caulk the joint into place. If the crack is crooked, chisel out the crack, clean loose debris, install a closed-cell foam, and caulk. This type of improvised control joint will prevent future cracking and keep the weather out.

EFFLORESCENCE

Efflorescence results when mineral salts leach out of masonry or mortar; both water and salts are necessary for efflorescence to occur. The basic cure for efflorescence is to shut off the source of the water and clean off the efflorescence. Water comes from condensation inside the building or from rain. Salts can come from masonry units, mortar, or other building elements. Salts can leach from backing materials to the facing wythe (see *Cleaning*, p. 187).

Sometimes further efflorescence can be prevented by sealing the wall to prevent water entry. However, sealing will cause further problems if the water continues to enter from somewhere, so it is vital to find the source of the problem first.

EXPANSION AND CONTRACTION

Masonry structures expand and contract due to changes in temperature, moisture content, and other chemical changes. However, masonry structures are rigid and unable to shift, and cracking often results from this movement. Cracked masonry is unsightly, hard to repair, weak, and vulnerable to water penetration.

The best prevention for such cracking is to allow the building to expand and contract by installing movement joints at the areas of greatest stress. Joint reinforcement, bond-beams, and grouted cavities also help deter cracking. In general, long walls and areas subject to temperature extremes or great stress need protection from cracking.

Cracking due to moisture loss (a factor for concrete units) can be prevented by ensuring that fresh units have about the same moisture content as they will have when the structure is in use (see *Joint Movement*, p. 244; and *Reinforcing*, p. 271).

SETTLING

Like most other masonry troubles, settling is easier to prevent than to fix. Almost every building settles to some extent, perhaps 3/8 inch to 1/2 inch. For this reason, reinforcing may be set in footings to prevent structural damage caused by uneven settling.

Stoops, step foundations, and additions that rest on poor footings or uncompacted subsoil are common places to find settling problems. These structures tend to settle away from the building.

Mudjacking (pumping grout under the footing) can be used to stabilize settling. In extreme cases, it may be necessary to rebuild the wall on a good footing. Brace the upper sections temporarily while doing this.

SPALLING

Spalling is a surface degradation of brick or block caused by freezing of water or by continuing efflorescence. The only way to repair a heavily-spalled surface to "like new" condition is to replace the units. Spalling is more often cleaned off and parged. To improve appearance, color the mortar to resemble the adjacent masonry. Waterproofing the surface may prevent further spalling, but you should try to remove the source of the water first.

STAINS

Stains are caused by external agents, such as smoke, welding spatter, oil, dirt and paint. Stains are different from efflorescence, which results when minerals leach from the masonry itself. Various products are on the market to remove stains. Always test the cleaning agent on a small spot before starting a large application (see *Cleaning*, p. 187).

WATER

Water causes many problems in masonry buildings. Running water can erode soil supporting the structure, causing a cave-in or settling. Water seeping under a slab can cause soil to expand and buckle the slab. Water freezing under footings can heave them, causing cracks and other damage.

Water that penetrates masonry units can cause other problems. It can start chemical reactions within some masonry elements, corrode lintels and other steel components, or damage related building elements, such as insulation and wallboard. The expansion of waterlogged units upon freezing can cause cracking.

Four major sources of moisture should concern masons:

- *Below-grade structures are exposed to "hydrostatic pressure" caused by the weight of water in the soil above. This pressure can be caused by a high water table or a heavy rain. Hydrostatic pressure causes the water to seek holes wherever it can. The best cures for hydrostatic pressure are prevention: keep water from accumulating near a foundation; provide drains so water can flow away. Existing hydrostatic pressure can be controlled by installing the preventative measures above, by patching leaks, by improving the gutters and downspouts, and by waterproofing the structure.*

- *Moisture can enter subgrade areas by capillary action (the tendency of water to flow through thin cracks or tubes). Surface treatments, such as a bituminous coating or bentonite, are used to prevent this capillary action.*

- *Rain is the principal source of moisture above grade. Proper joint tooling, flashing, coping, and coatings all help prevent rain from entering a structure.*

- *Moisture is also created by activities within a structure. Cooking, showering, and many commercial and industrial practices create humidity that must be allowed to escape.*

The best way to deal with moisture problems is to design and build properly to begin with. Repairs are difficult and expensive. In general, they parallel the proper practices described under *Coatings and Moisture Resistance*, p. 195; and *Flashing*, p. 231.

WEATHER

More than most trades, masons are at the mercy of the weather. This is because the properties of mortar and masonry units change along with temperatures. By choosing materials properly and using good work habits, the effects of hot and cold weather can be minimized.

COLD

Mortar behaves differently at low temperatures than at normal ones. Initial set, final set, and the development of strength are all delayed; less water is needed for a given level of plasticity; and air-entraining additives have less effect. Hydration below freezing is so slow that heat may be required before, during, and after laying units to ensure enough hydration.

Wind speed greatly influences the rate at which masonry units and mortar cool, so it is advisable, for the integrity of the job as well as the masons' comfort, to install windbreaks in cold, windy weather.

The following recommendations are taken from *Recommended Practices and Guide Specifications for Cold Weather Masonry Construction*, published by The International Masonry Industry All-Weather Council.

The Council reached a consensus that "masonry should be constructed in such a manner that it will develop sufficient strength and that the mortar will lose sufficient water to prevent expansion of the masonry upon freezing. Further, all masonry frozen during the early periods after construction should be moistened either naturally or artificially to reactivate the cement hydration process, which in turn will promote further strength development of the masonry."

Freezing can expand and displace the masonry units, so mortar should be prevented from freezing if possible. However, mortar with water content below 6 percent will not expand significantly upon freezing, so it is good practice to use dry mortar in cold weather.

TABLE 8.5
Possible Effects and Sources of Moisture Penetration

Sources of Moisture Penetration / Effects of Moisture Penetration	Previous Acid Cleaning *See Technical Notes 20 Revised*	Previous Sandblasting *See Technical Notes 20 Revised*	Plant Growth	Deteriorated Sealants/Caulks	Missing/Clogged Weepholes *See Technical Notes 21B*	Incompletely Filled Mortar Joints *See Technical Notes 7B Revised*	Capillary Rise	Broken/Loose Units	Differential Movement *See Technical Notes 18 Series*	Missing Flashing *See Technical Notes 7 Series*
Efflorescence *See TN 23 Series*	✓		✓	✓	✓	✓	✓	✓		✓
Deteriorated Mortar	✓	✓	✓			✓	✓	✓		
Spalled Units		✓		✓	✓	✓	✓		✓	✓
Cracked Units				✓	✓	✓	✓		✓	✓

TABLE 8.5 (cont.)
Possible Effects and Sources of Moisture Penetration

Effects of Moisture Penetration \ Sources of Moisture Penetration	Previous Acid Cleaning See Technical Notes 20 Revised	Previous Sandblasting See Technical Notes 20 Revised	Plant Growth	Deteriorated Sealants/Caulks	Missing/Clogged Weepholes See Technical Notes 21B	Incompletely Filled Mortar Joints See Technical Notes 7B Revised	Capillary Rise	Broken/Loose Units	Differential Movement See Technical Notes 18 Series	Missing Flashing See Technical Notes 7 Series
Rising Moisture					✓		✓			✓
Corrosion of Back-up Materials	✓			✓	✓	✓	✓	✓	✓	✓
Mildew/Algae Growth	✓			✓	✓	✓	✓	✓	✓	✓
Damaged Interior Finishes	✓			✓	✓	✓	✓	✓	✓	✓

Mortar that freezes briefly can be rescued by bringing it back to proper curing conditions quickly—mainly by heating and by adding water to allow the portland cement to hydrate properly. However, frozen mortar will probably never be as weather-resistant or watertight as mortar that cured properly.

Absorbent units absorb water from the mortar, reducing its average water content and susceptibility to freezing. Low-absorption units are difficult to lay in cold weather because they withdraw little water and tend to float. Thus, when working with low-absorption units, you may need auxiliary heat in conditions where you would not need it for laying absorbent units.

ADMIXTURES

Although several admixtures are used for cold weather masonry, the Council was negative toward most of them.

Antifreeze. Antifreeze is not recommended because it is only effective at concentrations that are strong enough to reduce the strength of the mortar. (Calcium chloride, commonly thought to be an antifreeze, is actually an accelerator—see below.)

Accelerators. Many accelerating chemicals are available, with the most common being calcium chloride. Calcium chloride is the major constituent of several proprietary accelerators and is also used under its own name. Calcium chloride causes corrosion in steel and should not be used if any reinforcement, wall ties, etc., are embedded in the wall. Calcium chloride can also cause efflorescence and excess shrinkage. If the wall contains no steel, use a maximum of 2 percent calcium chloride by the weight of the portland cement, or 1 percent by the weight of masonry cement.

Type III portland cement, high-early strength, is useful at low temperatures because it sets and achieves a strong bond faster than regular portland cement. Type M masonry cement has the highest portland cement content and the fastest set of the masonry cements.

Air-entraining additives. Air-entraining additives improve workability at cold temperatures, but they have not been proven effective at preventing damage from mortar freezing. Air-entraining additives are often blended directly in cement. Excessive air-entrainment will reduce compressive and bond strength and is not suggested, but cement containing air-entraining additives is allowed.

Coloring agents. Some coloring agents have a negative effect on cold-weather masonry. Carbon black should be limited to 2 percent of the cement content by weight. Dispersal agents used to distribute the pigment in some coloring agents may act as retarders and should not be used in cold weather.

Recommendations for cold weather. As a summary of safe, sound construction during cold weather, The International Masonry Industry All-Weather Council recommends:

1. Closely follow guidelines for cold-weather construction.
2. Store masonry units in a dry location. Thaw frozen sand and units.
3. Measure unit temperature if indicated. Units below 20°F must be heated but not overheated.
4. Heat mortar ingredients so the mortar temperature stays between 40°F and 120°F. Make every effort to have all batches fall within this range. Use a heater under the mortar board if needed.
5. Place masonry only on sound, unfrozen foundations. Never place it on a snow or ice-covered surface.
6. Cover work at the end of the day at least two feet down from the top.

The Council recommends the following specific procedures:

Air temperature	
Between 32°F and 40°F:	Heat sand *or* mixing water to between 40°F and 120°F.
Between 25°F and 32°F:	Heat sand *and* mixing water to between 40°F and 120°F.

Air temperature (cont.)	
Between 20°F and 25°F:	Heat sand *and* mixing water to between 40°F and 120°F. Use heat on both sides of walls under construction. Use windbreaks if wind is above 15 mph.
20°F and below:	Heat sand *and* mixing water to between 40°F and 120°F. Provide enclosure and auxiliary heat to maintain air temperature above 32°F. Temperature of units when laid shall be at least 20°F.

AFTER LAYING

Masonry must be protected after laying in cold weather. The Council recommends the following procedures:

Air temperature	
Between 20°F and 25°F:	Completely cover masonry with insulating blankets for 24 hours.
20°F and below:	Maintain masonry temperature above 32°F for 24 hours by enclosure and supplementary heat from electric blankets, heat lamps, or other approved method.

During cold weather, do not:

- *Use frozen mortar.*
- *Lay mortar on frost or ice.*
- *Use calcium chloride in masonry containing reinforcements, wall ties, or other steel.*

HEATING MATERIAL

Several types of heaters are used for heating sand. Tunnel-shaped stoves, sometimes built of culverts, can be heaped with sand. Steam heaters or propane-burning heaters are also used. Keep the sand moving to avoid scorching, which will reduce the strength of the resulting mortar.

Water can be heated by lighting fires underneath barrels, or by inserting steam probes or other immersion heaters into barrels. Try to attain a mortar temperature between 40°F and 120°F. Once the desired heat is reached, keep subsequent batches to the same temperature to simplify work.

Masonry units can be heated by setting up tents or storing them in finished, heated areas of the structure. Units must be heated if they are icy or below 20°F.

HOT

Masonry undergoes the following changes at high temperatures:

- *Mortar is less workable and more water is needed for a given level of workability.*
- *Units suck more water from the mortar, depending on unit characteristics, mortar and unit temperature, and unit moisture content.*
- *Air-entraining agents are less effective.*
- *Initial set, final set, and evaporation all occur faster.*

These changes add up to a mortar that is harder to work, faster to harden and set, and needs retempering more often. If the mortar loses too much water, it will not hydrate adequately. This is particularly likely to happen outdoors, where the rate of evaporation is much greater than indoors. Weak outer surfaces reduce the buckling strength of a wall, making it subject to damage from all kinds of loading.

Several practices will help counteract these hot weather changes:

- *Cool metal equipment that holds mortar. Keep wheelbarrows, mixing equipment, and mortar boards out of the sun or rinse them with cool water before they contact mortar.*
- *Store material in the shade. Cool the units by spraying with water when air temperature is above about 100°F (38°C). Allow units to dry before use. Note that such evaporative cooling is only effective when the relative humidity is low.*
- *Store sand in the shade. Sand at the job site usually contains water and natural evaporation will cool it. During hot conditions, spray the sand with water to maintain the cooling effect. As it evaporates, one gallon of water will cool one cubic yard of sand by 20°F.*
- *Use windscreens and fog sprays to reduce the drying effects of hot weather. Cover completed sections to slow the water loss and retain water for*

hydration. Damp curing is particularly valuable in sections that must have high tensile strength. Thin mortar or coatings which are particularly subject to drying.

- *Schedule construction to avoid the hottest part of the day.*
- *Use cool water for mixing mortar. Add ice to the mixing water if cool water is not available.*
- *Mix small batches of mortar to avoid rapid setting and excessive retempering.*
- *Work in the shade or erect a shade around the site.*

WINDOWS

Window openings are listed and detailed in blueprints. In general, they are similar to doors. Movement joints are often located at windows because walls are likely to crack near openings. If a window is wider than 6 feet, it is advisable to use one movement joint on either side.

Window openings may be listed as rough opening dimensions on blueprints. Opening locations and sizes should be planned to conform to the building module to minimize cutting of units. When stringing units dry, try to make the layout match the rough opening dimensions. Sometimes nonmodular windows can be made to fit modular units by slightly changing the size of head joints. If units must be cut, make sure not to cut jamb units smaller than 4 inches, because they will not be strong enough to bear the weight of the lintel.

Any vertical adjustments needed to make the courses match the window top should be made by varying the spacing of courses, not by installing a course of splits. The course height must be maintained throughout the building, another reason for using modular windows.

Masons are given the following dimensions for a window installation: location in the wall; height, width, and jamb depth of the window; and height, width, and thickness of the rough opening. Sometimes, windows are already in place when masons begin work. Consult the window schedule (each window has a letter on the blueprint corresponding to an item on the schedule). Also check the window manufacturer's directions before laying out windows.

SETTING A WINDOW
IN A VENEER WALL

The following procedure (see Fig. 8.47) uses most of the techniques of laying a veneer wall with the special requirements of building a rough opening around a window in a wood- or steel-studded wall. The foundation, brick ledge, and window are already in place (see *Walls, veneer* for details on weepholes and flashing, p. 125):

1. Lay out the wall. Measure the actual wall, or read the plans. Measure up from the brick ledge to the top of the window sill. Subtract the height of the masonry sill from this height to find the "height to the sill course."

2. Figure the joint spacing. Use a brick spacing rule or a modular rule to find a course spacing that will reach the height to the sill course with only whole courses. Mark the courses on the story pole.

3. Check that this coursing will reach all sills around the building. All courses must be even, and no splits are to be used below sills.

4. String units dry along the wall and determine a joint spacing that will minimize cutting. Write down the dimensions of any units that must be cut and mark the head joints on the foundation.

5. Determine the spacing between the backing wall and veneer—generally about 3/4 inch to 1 inch. Insulate this space if required.

6. Mark the exterior line of the veneer on the brick ledge with a chalk line if the ledge is wide enough.

7. Lay the first course. Install flashing at the wall base, using mastic or mortar as required. Fasten the inside of the flashing to the backing wall so the flashing slopes to the outside.

8. Make weepholes at the proper spacing (generally 16 inches to 24 inches O.C.)

9. Anchor the wall with wall ties. Use ties about every five courses, with extra ties near windows and doors and in panels

above 20 feet high. Wall ties must have at least a 2-inch purchase in the bed joints.

10. Construct a rowlock sill at the proper height. Mark the horizontal extent of the sill by plumbing down from the outside trim of the window and marking on the bricks. Use a brick spacing rule to determine jointing for the sill. If you are insulating the cavity, make sure to pour insulation into the cavity beneath the window before placing the rowlock course.

11. Sills should protrude about 3/4 inch to 1 inch and slope to the outside for drainage. Figure the length of the sill bricks and the angle of the cut. Cut the proper number of sill bricks.

12. Install a flashing into the bottom of the sill course and nail it to the framing.

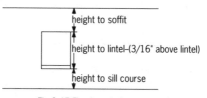

height to soffit

height to lintel—(3/16" above lintel)

height to sill course

Fig.8.47 Figuring window courses

13. Lay a beveled mortar bed across the sill width. Begin laying the bricks into position, using the joint spacing figured above. The rowlocks should slope down at the correct angle and the course should be level from side to side (see Fig. 8.48).

14. Figure the height from the sill course (the course supporting the rowlock course) to the lintel height (3/16 inch above the top of the window frame). This lintel height will allow a tight caulking seal between lintel and head jamb and prevent the lintel from bearing upon the head jamb. Use the brick spacing rule to figure proper course height between the sill course and the lintel height.

15. Check that this joint spacing will work with all other windows, then mark it on the story pole. Continue laying the jamb courses, affixing wall ties as needed.

16. When lintel height is reached, check that both bearing surfaces are level. Install the lintel with proper bearing (at least 4 inches) so it is level, square, and gives proper clearance for the veneer.

17. Clip the back of bricks if needed to let them rest solidly on the lintel. Determine the course spacing required to reach the soffit. Lay the wall to the soffit and caulk the joint at the soffit.

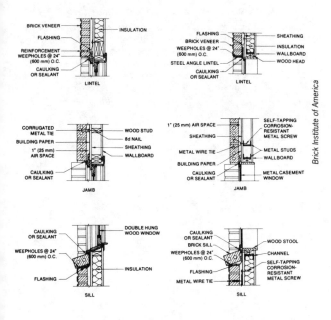

Fig. 8.48 Lintel, jamb and sill details

VENTS AND FANS

Vents and fans are used in walls to provide air circulation and to allow moisture to exit a structure. Metal louvers keep rain out of the opening, and screens prevent insects and animals from entering. Most vents and fans are made to fit with modular construction with little or no cutting. Other trades may be responsible for placing a vent in an opening left by a mason.

TYPES OF VENTS

- *A block can be turned on its side to provide ventilation through the cores.*

- *Ready-made flat vents, usually 1 inch × 8 inches × 16 inches, are used for ventilating foundation courses. The vents are usually placed in the top course of the foundation below the plate. Some vents slip into grooves in sash blocks. Wire mesh can be used to cover cores of the blocks below the vent. Cover the mesh with mortar to create a sill so water can run off. Tuckpoint or caulk the sash block grooves after the vent is placed.*

- *Other vents and fans may be specified by the architect.*

PART FIVE

SOME FUNDAMENTALS

9

Estimating Jobs

Estimating is one of the toughest challenges facing a masonry contractor. Estimate too low, and you can lose money. Estimate too high, and your bid can be rejected.

Contractors must know how to estimate both materials and labor. Formulas given in books on construction estimating can help you estimate labor and material requirements, and provide total costs for specific wall types based on:

1. The amount of structure to be built (for example, 100 square feet of veneer wall with standard modular brick requires x man-hours).

2. The number of units to be laid (for example, laying 1000 8 × 8 × 16 concrete blocks requires x man-hours).

Estimating books may state that the job must be performed under certain conditions. For example, 100 square feet of modular brick veneer requires x man-hours on a wall less than three stories high. Estimating books are only helpful if workers meet standard rates of production and local costs are similar to those used by the organization preparing the book (see Table 9.1).

Contractors also use their own records for estimating jobs, which is a good reason to keep records up to date.

Follow these hints for accurate estimating:

* *Keep a record of time requirements and cost per unit of masonry and per square foot for past jobs.*

* *Add at least 5 percent to brick and block materials estimates to allow for mistakes and breakage.*

309

- *Add at least 10 and 25 percent to mortar estimates for waste.*

- *Complicated structures, such as fireplaces and chimneys, require more labor per unit of material than simpler structures.*

- *When ordering materials, remember that special units may be required for bond-beams, lintels, half-height courses, grouted cores in high-lift grouting, etc. If a significant number of special units will be used, reduce the number of standard units accordingly.*

FIGURING WALL AREA

Wall area can be calculated by finding the total number of square feet of each wall and subtracting the area of openings (see Fig. 9.1). For example, find the area of a wall 42 feet long × 8 feet high with three windows, each 2 feet × 4 feet and one door, 3 feet × 7 feet. When calculating wall area for estimating, use rough-opening (RO) dimensions:

1. Gross area (42 feet × 8 feet) =   336 square feet
2. Minus windows (2 feet × 4 feet) =
   8 square feet each × 3 windows =   −24 square feet
3. Minus door 3 feet × 7 feet   <u>−21 square feet</u>
4. Net area   291 square feet

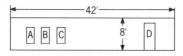

Windows A, B, And C = 2' X 4'
Door D = 3' X 7'

Fig. 9.1 Calculating wall area

BRICKS

Once the square footage is figured, you can use three techniques for estimating materials:

1. Use Table 9.2 to find the number of bricks per 100 square feet. If the bricks have a nominal size of 4 inches × 2-2/3 inches × 8 inches, 675 bricks are required per 100 square feet.

TABLE 9.1
Masonry Estimator

Brick	Face brick, 4" (at $260.00 per 1,000) up to 8' high	Unit	Material	Labor	Total
veneer, single wythe, running bond, 3/8" joint	Commercial	SF	2.55	4.85	7.40
	School or institutional	SF	2.65	5.00	7.65
	Modular	SF	2.55	4.85	7.40
	Norman brick	SF	2.76	5.05	7.81
	Roman brick	SF	3.48	6.95	10.43
	Glazed brick, standard size	SF	5.10	10.00	15.10
Brick walls, common brick, running bond	8" wall	SF	3.85	7.05	10.90
	12" wall	SF	5.82	10.50	16.32
Add for:	American bond	%	—	—	15.0
	Basketweave bond	%	—	—	20.0
	Flemish bond	%	—	—	25.0
	Herringbone pattern	%	—	—	30.0
	Soldier bond	%	—	—	20.0
	Stacked bond	%	—	—	15.0
	Institutional inspection	%	—	—	20.0
	Short runs or cut-up work	%	—	—	20.0
	Pilasters to 15 SF	%	—	—	100.00
	Walls over 1 story, per story	%	—	—	10.0
	Modular brick	%	—	—	15.0
	Colored mortar, red, tan, brown	SF	.25	—	—
Clay brick specialties	Arches	LF	—	—	17.90
	Caps and coping	LF	—	—	7.60
	Cuts	LF	—	—	5.50
	Heads, anchored	SF	—	—	73.00
	Jambs	LF	—	—	6.10
	Dove tail anchors	Ea	.25	.45	.70
Mortar filling for 2" wall cavity, type S mortar at $2.50 per CF, per SF of wall face		SF	.50	.46	.96
Zonolite fill for 2" cavity wall		SF	.22	.27	.49

NATIONAL CONSTRUCTION ESTIMATOR 1990
CRAFTSMAN BOOK CO., P.O. BOX 6500
CARLSBAD, CA 92008

291 square feet ÷ 100 square feet × 675 bricks = 1964 bricks
Finally, add 5 percent for waste. This can be done simply by
multiplying by 105 percent (1.05)
1964 bricks × 1.05 = 2062 bricks required.

2. An alternate method is to find the square inches of wall each unit
   of the nominal size occupies, then find how many units per
   square foot. Multiply by the total square footage to find the
   number of units required.

   2-2/3 inches × 8 inches = 21.28 square inches per brick.
   144 square inches per square foot ÷ 21.28 = 6.77 bricks per
   square foot
   6.77 × 291 = 1970 bricks required.
   Then add 5 percent for waste:
   1.05 × 1970 = 2068.50 bricks required.

   If the wall specifies two wythes of the same brick, multiply the
   requirement by two. For bonds other than running bond, a
   larger number of face brick are needed. Use the correction
   factors shown in Table 9.2 to increase the face brick quantity.
   Reduce the backing wythe brick quantity by the same amount.

3. Another approach to these calculations is to use a masonry
   estimator, such as the one developed by the Brick Institute of
   America. This estimator will figure wall area from length and
   width, calculate the number of bricks (of several sizes) required,
   and the cost of that brick. It will also figure mortar for head and
   bed joints, and for collar joints. Finally, the estimator will figure
   sand, cement, and lime requirements for the mortar. To make
   all these calculations, you must know the size and cost of the
   brick, the dimesions of the wall, and the size of the head, bed, and
   collar joints. The estimator does not allow for waste or openings.

MORTAR

The mortar requirement depends on the joint size. The easiest
method is to use a chart or a masonry estimator. Mortar can be
figured per 100 square feet, per 1000 units, or by the number of units
that can be laid with one bag of cement. Allow at least 10 to 25
percent extra for mortar, to allow for waste and variations in unit
size. In hot weather, which speeds up the set of mortar, you might

TABLE 9.2
Brick and Mortar Requirements
Nonmodular brick and mortar required
for single-wythe walls in running bond

Size of Brick in.			With ⅜-in. Joints		
t	h	l	Number of Bricks per 100 Sq Ft	Cubic Feet of Mortar per 100 Sq Ft	Cubic Feet of Mortar per 1000 Brick
2¾ × 2¼ × 9¾			455	3.2	7.1
2⅝ × 2¾ × 8¾			504	3.4	6.8
3¾ × 2¼ × 8			655	5.8	8.8
3¾ × 2¾ × 8			551	5.0	9.1

Size of Brick in.			With ½-in. Joints		
t	h	l	Number of Bricks per 100 Sq Ft	Cubic Feet of Mortar per 100 Sq Ft	Cubic Feet of Mortar per 1000 Brick
2¾ × 2¼ × 9¾			432	4.5	10.4
2⅝ × 2¾ × 8¾			470	4.1	8.7
3¾ × 2¼ × 8			616	7.2	11.7
3¾ × 2¾ × 8			522	6.4	12.2

TABLE 9.2 (cont.)
Bond correction factors for walls
(Add to facing and deduct from backing.)

Bond	Correction Factor[1]
Full headers every 5th course only	1/5
Full headers every 6th course only	1/6
Full headers every 7th course only	1/7
English bond (full headers every 2nd course)	1/2
Flemish bond (alternate full headers and stretchers every course)	1/3
Flemish headers every 6th course	1/18
Flemish cross bond (Flemish headers every 2nd course)	1/6
Double-stretcher, garden wall bond	1/5
Triple-stretcher, garden wall bond	1/7

[1] CORRECTION FACTORS ARE APPLICABLE ONLY TO THOSE BRICKS WHICH HAVE LENGTHS OF TWICE THEIR BED DEPTHS.

TABLE 9.2 (cont.)
Modular brick and mortar required for single-wythe
walls in running bond (no allowances for breakage or waste).

Nominal Size of Brick in.			Number of Brick per 100 sq ft	Cubic Feet of Mortar			
				Per 100 Sq Ft		Per 1000 Brick	
t	h	l		⅜ - in. Joints	½ - in. Joints	⅜ - in. Joints	½ - in. Joints
4 ×	2⅔ ×	8	675	5.5	7.0	8.1	10.3
4 ×	3⅕ ×	8	563	4.8	6.1	8.6	10.9
4 ×	4 ×	8	450	4.2	5.3	9.2	11.7
4 ×	5⅓ ×	8	338	3.5	4.4	10.2	12.9
4 ×	2 ×	12	600	6.5	8.2	10.8	13.7
4 ×	2⅔ ×	12	450	5.1	6.5	11.3	14.4
4 ×	3⅕ ×	12	375	4.4	5.6	11.7	14.9
4 ×	4 ×	12	300	3.7	4.8	12.3	15.7
4 ×	5⅓ ×	12	225	3.0	3.9	13.4	17.1
6 ×	2⅔ ×	12	450	7.9	10.2	17.5	22.6
6 ×	3⅕ ×	12	375	6.8	8.8	18.1	23.4
6 ×	4 ×	12	300	5.6	7.4	19.1	24.7

COURTESY OF THE BRICK INSTITUTE OF AMERICA

want to increase the waste factor. Some contractors figure waste as high as 50 or 100 percent of the basic mortar requirement.

COLLAR JOINTS

The Brick Institute of America recommends the following mortar requirements for collar joints (per 100 square feet of wall):

1/4-inch joint	2.08 cubic feet
3/8-inch joint	3.13 cubic feet
1/2-inch joint	4.17 cubic feet

For thicker collar joints, divide the actual collar joint by one of the listed collar joints, then multiply by the mortar requirement. For example, a 1-1/4-inch collar joint would be figured like this:

5/4 ÷ 1/4 = 5; 5 × 2.08 cubic feet per 100 square feet of wall = 10.40 cubic feet.

BLOCK

The calculation for block required is similar to that for brick. Two methods are possible—figuring courses and blocks per course, or figuring square footage:

1. Assume the same wall as above: 42 feet × 8 feet with 45 square feet of openings.

2. Because most concrete masonry units have a nominal length of 16 inches, each block is 16/12 feet, or 1.33 feet long. Divide wall length by 1.33 to find blocks per course.

   42 feet ÷ 1.33 feet/block = 31.5 blocks per course

3. Each course is 8 inches high, which equals 2/3 foot. Thus 1-1/2 courses are required per foot of height. To find courses needed, multiply the wall height by 1.5:

   8 feet × 1.5 courses/foot = 12 courses

4. Multiply blocks per course by number of courses to find gross blocks needed:

   31.5 blocks per course × 12 courses = 378 blocks

5. Now adjust for openings by calculating their area, as above. Note that each 8-inch × 16-inch block occupies 128 square inches. Dividing 128 square inches by 144 square inches per square foot shows that each block occupies .89 square feet of wall.

   45 square feet ÷ 0.89 square feet per block = 50.5 blocks
   (use 50)

6. Subtract blocks not needed for openings from gross blocks to find net blocks required:

   378 – 50 = 328

7. Then add the 5 percent waste factor.

   32 × 1.05 = 344.4 blocks required (use 345).

A simpler alternative is to calculate gross square footage, subtract area of openings, and multiply by blocks per square foot:

1. Gross area (8 feet × 42 feet) =	336 square feet
2. Minus openings (from above) =	– 45 square feet
3. Net wall area =	291 square feet

4. Find the number of blocks:

   291 square feet ÷ 0.89 = 326.97 blocks (use 327)

5. Add 5 percent for waste:

   327 × 1.05 = 343.35 blocks required (use 344).

Tables can also be used for estimating block requirements. In multiwythe walls, block requirements depend on the materials chosen for each wythe and the method of tying the wythes together.

TABLE 9.3
Concrete Masonry Unit Requirements

Description, Size of Block (in.)	Thickness Wall (in.)	Weight Per Unit (lb.)	Number of Units per 100 Sq. Ft. of Wall Area	Mortar (cu. ft.)	Weight, Pounds per 100 Sq. Ft. of Wall Area
8 × 8 × 16	8	50	110	3.25	5860
8 × 8 × 12	8	38	146	3.5	6000
8 × 12 × 16	12	85	110	3.25	9700
8 × 3 × 16	3	20	110	2.75	2600
9 × 3 × 18	3	26	87	2.5	2500
12 × 3 × 12	3	23	100	2.5	2550
8 × 3 × 12	3	15	146	3.5	2550
8 × 4 × 16	4	28	110	3.25	3450
9 × 4 × 18	4	35	87	3.25	3350
12 × 4 × 12	4	31	100	3.25	3450
8 × 4 × 12	4	21	146	4	3500
8 × 6 × 16	6	42	110	3.25	5000

COURTESY OF THE BRICK INSTITUTE OF AMERICA

STONE

Stone is sold by the ton. Suppliers have estimates for square foot coverage per ton, so it is simple to figure material requirements. Always allow at least 10 percent for waste.

Four-inch thick veneers usually cover about 40 square feet per ton. Lighter stone, such as lava, may cover up to 50 to 80 square feet per ton. Many building stones weigh about 135 to 180 pounds per cubic foot.

WALL TIES AND REINFORCEMENT

If wall ties are to be used, divide wall area by the number of ties needed per square foot.

Assume the above wall, with 291 square feet net area. If one wall tie is required per 4.5 square feet, then:

291 square feet ÷ 4.5 square feet per tie = 64.7 ties. Add a few extra ties for areas near openings.

The amount of joint reinforcement depends on the linear feet of courses and the vertical spacing. Assume a wall 32 feet long and 8 feet high (12 courses high) will be reinforced every other course (16 inches on center). This means that 6 courses must be reinforced. Multiply the wall length by the number of courses to reinforce:

32 feet × 6 = 192 feet.

An easier method for walls that will be reinforced 16 inches on center is to multiply the wall length by half the number of courses:

32 feet × 12/2 = 192 feet.

After figuring reinforcement, add 10 percent for overlap and waste. Be sure to specify the proper corners and tee intersections if these will be purchased prefabricated. If not, increase the waste figure.

The quantity of reinforcing rod needed is the sum of the rods needed for bond beams and for grouted cores:

1. Check specifications for the spacing of the bond-beams. Figure how much material is required for each bond-beam, then multiply by the number of bond-beams needed.

2. Find the rod required for each core, then multiply by the number of cores to be reinforced. If the walls are higher than 8 feet, allow extra for overlapping of the bars. Thirty times the rod diameter is the standard overlap.

10
Mathematics for Masons

Masons must be able to perform basic arithmetic of feet, inches, and fractions. They must also know how to measure angles and calculate area and volume.

MEASURING ANGLES

Angles are measured with a system of degrees. 360 degrees (written 360°) equals one full circle. 90° is a right angle and 180° is a straight angle. Degrees are broken into 60 minutes (written 60'). Each minute is broken into 60 seconds (written 60"), but masons rarely need to deal with the accuracy represented by seconds.

Angles less than 90° are called acute angles. Angles between 90° and 180° are called obtuse angles.

AREA, PERIMETER, AND VOLUME OF SHAPES

Perimeter is the linear dimension of the outside of a two-dimensional shape. Perimeter is useful for calculating material needs of entire structures. Area is the amount of space occupied by a two-dimensional shape. Area calculations are needed when estimating material requirements for walls.

VOLUME OF SHAPES

Bricklayers must also know how to figure volume of shapes. Volume is the three-dimensional space occupied by a structure. Volume is useful in calculating the weight of structures and in figuring grout and concrete needs (see Fig. 10.1).

Shape	Area	Perimeter	Volume
Triangle	$B \times H \div 2$	$AB + CB + AC$	
Rectangle	$L \times W$	$2 \times (L + W)$	
Circle	$\pi \times R^2$	$\pi \times D$	
Sphere	$4 \times \pi \times R^2$		$4/3 \times \pi \times R^3$
Cylinder	$(2\pi \times R^2) + (\pi \times D \times H)$		$\pi \times R^2 \times H$

area of triangle = 1/2BH

area of rectangle = LW

π = 3.14

area of circle = πR

area of cube, rectangular prism or column = LWH

Fig. 10.1 Area of shapes

Notes: Circumference is the term used for the perimeter of a circle. R^2 means "radius squared." To square a number, multiply it by itself. π, pronounced "pie," is the ratio of circumference to diameter. 3.14 is close enough to the actual value of π for most masonry work.

$$\text{Diameter} = \text{Radius} \times 2$$
$$\text{Radius} = \text{Diameter} \div 2$$

AREA OF AN IRREGULAR SHAPE

To find the area of an irregular shape, try to break the shape down into components. Thus, the simple house plan in Fig. 10.2 would be broken down into the main section, porch, and stoop.

Main section	=	42 feet × 28 feet	=	1176 square feet
Porch	=	14 feet × 18 feet	=	252 square feet
Stoop	=	4 feet × 8 feet	=	32 square feet
Total	=			1460 square feet

To find the net area of a wall with openings, see *Estimating Jobs*, p. 309.

AREA IN FEET AND INCHES

If feet and inches are used to describe the length and width of buildings, you must first change the inches to fractions of a foot

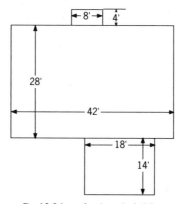

Fig. 10.2 Area of an irregular building

before multiplying. Ignore fractions of an inch for most material estimates. Example: Find the area of a rectangle 6 feet 5 inches wide by 12 feet 4 inches

1. Convert the inches to feet by dividing by 12:
   5 inches = 5/12 feet, so 6 feet 5 inches = 6-5/12 feet
   4 inches = 4/12 feet, so 12 feet 4 inches = 12-4/12 feet

2. Convert these measurements to improper fractions:
   6-5/12 feet = 77/12 feet
   12 4/12 feet = 148/12 feet

3. Find the area by multiplying width times length
   77/12 feet × 148/12 feet = 11396/144 square feet

4. Reduce this fraction by dividing numerator by denominator:
   11396 ÷ 144 = 79.14 square feet. This is the area of a rectangle 6 feet 5 inches wide by 12 feet 4 inches long.

This calculation could be simplified by rounding 6 feet 5 inches to 6.5 feet and 12 feet 4 inches to 12.3 feet. Then multiply on paper or with a hand calculator:

6.5 × 12.3 = 79.9 square feet.

This is usually adequate for estimates.

AREA AND VOLUME CONVERSIONS

1 square inch = 1/144 square foot
1 square foot = 144 square inches
1 square yard = 1296 square inches
1 square foot = 1/9 square yard
1728 cubic inches = 1 cubic foot
1 cubic inch = 1/1728 (0.00057) cubic foot
1 cubic foot = 1/27 (0.037) cubic yard
1 cubic yard = 27 cubic feet

AREA IN ENGINEER'S MEASUREMENT

Finding wall area from plans that are written in engineer's measurement is simple if a hand calculator is available.

▪ EXAMPLE:

What is the area of a wall 98.6 feet by 8.66 feet?

98.6 × 8.66 = 853.88 square feet
(This can be rounded to 854 square feet for most purposes.)

11

Measurements

Measurements are part of the daily life of a mason. Measuring is used to translate figures on blueprints into structures that faithfully mirror the blueprints. Because accuracy is essential, thorough knowledge of measuring systems is important.

UNITED STATES SYSTEM

The measuring system now used in the United States was once common in other countries. Now, these other nations have switched to the metric system, leaving the United States the only major country using feet and inches. In this system, twelve inches equal one foot. Inches are usually divided into fractions. If an inch is divided into eight equal parts, then each part is called "one-eighth," which is written 1/8. Two of these units, 2/8, equals 1/4 because you can "simplify" the fraction.

ENGINEER'S MEASUREMENT

Engineers and architects use a system called "engineer's measurement." In this system, distances are expressed in feet and tenths, hundredths, and thousandths of a foot: 12 feet 6 inches would be expressed as 12.5 feet (inches are not used in this system). Masons must be able to convert engineer's measurement to feet, inches, and fractions of inches. The easiest way to do this is by using a table, such as Table 11.1. Notice also where each inch falls in the engineer's measurement system, that each 0.01 foot is very close to 1/8 inch, and that 0.04 foot is very close to 4/8 inch or 1/2 inch.

METRIC—UNITED STATES CONVERSIONS

Metric measurements are slowly growing more common in the construction trades, and masons must know how to deal with them. The metric system uses decimals instead of fractions. This system may be unfamiliar, but it is easy to use because fractions play no part, and decimal calculations are easy with a hand calculator.

TABLE 11.1
Engineer's measurements

Decimal Feet	Inches	Fractions	Decimal Feet	Inches	Fractions	Decimal Feet	Inches	Fractions
0.01		1/8"	0.24	2	7/8"	0.47	5	5/8"
0.02		1/4"	0.25	3"		0.48	5	3/4"
0.03		3/8"	0.26	3	1/8"	0.49	5	7/8"
0.04		1/2"	0.27	3	1/4"	0.50	6"	
0.05		5/8"	0.28	3	3/8"	0.51	6	1/8"
0.06		3/4"	0.29	3	1/2"	0.52	6	1/4"
0.07		7/8"	0.30	3	5/8"	0.53	6	3/8"
0.08	1"		0.31	3	3/4"	0.54	6	1/2"
0.09	1	1/8"	0.32	3	7/8"	0.55	6	5/8"
0.10	1	1/4"	0.33	4"		0.56	6	3/4"
0.11	1	3/8"	0.34	4	1/8"	0.57	6	7/8"
0.12	1	1/2"	0.35	4	1/4"	0.58	7"	
0.13	1	5/8"	0.36	4	3/8"	0.59	7	1/8"
0.14	1	3/4"	0.37	4	1/2"	0.60	7	1/4"
0.15	1	7/8"	0.38	4	5/8"	0.61	7	3/8"
0.16	1	15/16"	0.39	4	3/4"	0.62	7	1/2"
0.17	2"		0.40	4	7/8"	0.63	7	5/8"
0.18	2	1/8"	0.41	4	15/16"	0.64	7	3/4"
0.19	2	1/4"	0.42	5"		0.65	7	7/8"
0.20	2	3/8"	0.43	5	1/8"	0.66	7	15/16"
0.21	2	1/2"	0.44	5	1/4"	0.67	8"	
0.22	2	5/8"	0.45	5	3/8"	0.68	8	1/4"
0.23	2	3/4"	0.46	5	1/2"	0.69	8	1/4"

TABLE 11.1 (cont.)
Engineer's measurements

Decimal Feet	Inches	Fractions	Decimal Feet	Inches	Fractions	Decimal Feet	Inches	Fractions
0.70	8	3/8"	0.81	9	3/4 "	0.91	10	15/16"
0.71	8	1/2"	0.82	9	7/8"	0.92	11"	
0.72	8	5/8"	0.83	10"		0.93	11	1/8"
0.73	8	3/4"	0.84	10	1/8"	0.94	11	1/4"
0.74	8	7/8"	0.85	10	1/4"	0.95	11	3/8"
0.75	9"		0.86	10	3/8"	0.96	11	1/2"
0.76	9	1/8"	0.87	10	1/2"	0.97	11	5/8"
0.77	9	1/4"	0.88	10	5/8"	0.98	11	3/4"
0.78	9	3/8"	0.89	10	3/4"	0.99	11	7/8"
0.79	9	1/2"	0.90	10	7/8"	1.00	12"	
0.80	9	5/8"						

In the metric system, the basic unit of length is the meter (39.37 inches). The basic unit of weight is the kilogram (2.2 pound). The metric system uses certain prefixes to indicate multiples of the basic units. The important prefixes are:

$$milli = .001$$
$$centi = .01$$
$$kilo = 1000$$

Thus one centimeter equals 0.01 meter, and one kilometer equals 1,000 meters. To convert from one measurement to another, just shift the decimal place. Thus 2.566 meters can be expressed as two meters 566 millimeters, 256.6 centimeters, or 2566 millimeters.

The metric system uses the Celsius degree of heat (formerly called centigrade). Each degree Celsius equals 1.8°F. Boiling is 100°C and freezing is 0°C in Celsius. See Table 11.2.

TABLE 11.2
Metric and American conversions

American	Metric
1 inch	2.54 centimeters
1 foot	30.4 centimeters
1 yard	91.4 centimeters
0.032 foot	1 centimeter
3.28 feet	1 meter
1.093 yard	1 meter
1 square foot	0.092 square meters (m²)
1 square yard	0.836 square meters
0.155 square inches	1 square centimeter (cm²)
10.763 square feet	1 square meter
1.195 square yard	1 square meter
1 cubic inch	16.387 cubic centimeters
1 cubic foot	0.028 cubic meters (m³)
1 cubic yard	0.764 cubic meters
0.061 cubic inch	1 cubic centimeter (cm³)
35.314 cubic feet	1 cubic meter
1.307 cubic yards	1 cubic meter

12
Safety

Construction is a dangerous occupation, and workers and management are obliged to help ensure a safe workplace. The following information is intended as minimum requirements and not as a comprehensive guide to safety on the job.

Some conditions greatly increase the chance for accidents in any working environment. These conditions include crowded work areas, blocked passageways, overloaded platforms and hoists, inadequate support for scaffolds, elevators without proper guards, damaged tools or power cords, poor ventilation or lighting, defective or worn out protective equipment and clothing, and improper storage or use of chemicals.

Some types of behavior also increase the odds of accident and injury, including poor attention to the task (especially when working with power equipment), creating sparks or flames near flammable materials, creating toxic fumes in unventilated areas, riding equipment not designed for personnel, removing guards from machinery, stacking materials unsafely, and neglecting to wear protective clothing when appropriate. Other dangerous habits include enagaging in horseplay and working when very tired.

For further information on tool safety, see *Part One: Tools*, p. 1.

The Occupational Safety and Health Administration (OSHA) is responsible for establishing and enforcing health and safety standards at workplaces. Although the detailed regulations are complex, some OSHA booklets summarize standards that are frequently overlooked or would prevent the most hazardous situations.

OSHA booklet number and name

2201 General Industry Digest
2202 Construction Industry
2226 Excavating and Trenching Operations
3097 Electrical Standards for Construction
3106 Concrete and Masonry Construction

OSHA has a free "Onsite Consultation Program" available to all employers. This program provides workplace inspections to recognize and correct hazards, and does not issue citations. See Table 12.1. The program is primarily targeted to small businesses which want to fulfill their requirements under the OSHA law. The employer's only obligation is a commitment to correct serious hazards. The following are some of the OSHA standards that apply to masonry construction:

ACCESS ZONE

Before wall construction starts, OSHA requires a zone be established to prevent injury from falling objects. This limited access zone must be

- *4 feet taller than the wall to be constructed;*
- *as long as the wall;*
- *on the side of the wall without scaffolding;*
- *restricted to employees actively engaged in constructing the wall; and*
- *in place until the wall is supported well enough to prevent overturning and collapse.*

In addition, unsupported walls taller than 8 feet must be braced until permanent supporting elements of the structure are in place.

ACCIDENT REPORTING

The employer must maintain a log and summary (OSHA Form no. 200 or equivalent) of all recordable injuries and illnesses for each work site. Reportable events are those which result in fatality, hospitalization, lost workdays, medical treatment, job transfer or termination, or loss of consciousness. Incidents must be entered in the log within six days after the employer learns of them. An annual summary of this log must be compiled each year and posted at the work site from February 1 until March 1.

DISPOSAL CHUTE

An enclosed chute must be used whenever materials are dropped more than 20 feet to any exterior point of a building. When debris is dropped through a hole in the floor, the area where the material

TABLE 12.1
OSHA Offices
U.S. Department of Labor Regional Offices
for the Occupational Safety and Health Administration

Region I	Region VI
(CT, MA, ME, NH, RI, VT) 16-18 North Street 1 Dock Square Building 4th Floor Boston, MA 02109 Telephone: (617) 223-6710	(AR, LA, NM, OK, TX) 555 Griffin Square Bldg. Room 602 Dallas, TX 75202 Telephone: (214) 767-4731
Region II	**Region VII**
(NJ, NY, Puerto Rico, Virgin Islands) 1 Astor Plaza, Room 3445 1515 Broadway New York, NY 10036 Telephone: (212) 944-3426	(IA, KS, MO, NE) 911 Walnut Street, Room 406 Kansas City, MO 64106 Telephone: (816) 374-5861
Region III	**Region VIII**
(DC, DE, MD, PA, VA, WV) Gateway Building, Suite 2100 3535 Market Street Philadelphia, PA 19104 Telephone: (215) 596-1201	(CO, MT, ND, SD, UT, WY) Federal Building, Room 1554 1961 Stout Street Denver, CO 80294 Telephone: (303) 837-3061
Region IV	**Region IX**
(AL, FL, GA, KY, MS, NC, SC, TN) 1375 Peachtree Street, N.E. Suite 587 Atlanta, GA 30367 Telephone: (404) 881-3573	(AZ, CA, HI, NV, American Samoa Guam, Pacific Trust Territories) Box 36017 450 Golden Gate Avenue San Francisco, CA 94102 Telephone: (415) 556-7260
Region V	**Region X**
(IL, IN, MI, MN, OH, WI) 230 South Dearborn Street 32nd Floor, Room 3244 Chicago, IL 60604 Telephone: (312) 353-2220	(AK, ID, OR, WA) Federal Office Building Room 6003 909 First Avenue Seattle, WA 98174 Telephone: (206) 442-5930

is dropped must be enclosed by barricades at least 42 inches high and at least 6 feet back from the edge of the opening. Signs warning of the hazard of falling material must be posted near where the debris falls.

ELECTRICITY

OSHA requires compliance with the 1971 National Electric Code in most cases. Circuits of 15 and 20 amp, single-phase, 120 volt current at construction sites must use ground fault interrupters or an assured equipment grounding program unless they are part of permanent structural wiring. One or more employees at the job site should make daily inspections of temporary wiring and electrical equipment.

Other guidelines for electrical safety:

1. Check electrical equipment before using. Equipment must be grounded. Don't use if it has frayed cords, damaged insulation, or exposed wires.

2. Never operate electrical equipment when standing in wet or damp areas.

3. Electric cords must be three-wire type. Do not hang cords from nails or wires. Splices must be soldered and adequately insulated.

4. Keep electrical wires off the ground. Never run over wires with equipment.

5. Plug receptacles must be the approved, concealed contact type.

6. Exposed metal parts of electrical tools that do not carry current must be grounded.

7. Shut off power in case of a problem or accident.

8. Do not touch anyone who is in contact with live electrical current. Instead, first shut off the power. Then move the person by pushing him or her with a dry piece of lumber. Give mouth-to-mouth resuscitation if the victim is not breathing. Call an ambulance.

FIRE

On larger jobs, a firefighting program must be followed throughout the construction work. The job site must have an alarm system to alert employees and the fire department of a fire.

Work sites must follow these fire prevention measures: Store oily or greasy rags in a tightly-closing metal container. Remove all flammable waste daily. Maintain adequate clearance between heat sources and flammable materials. Keep salamanders and welding operations at least 10 feet from canvas or plastic tarpaulins. (See below for rules on flammable liquids.)

Each floor of a building must have two separate fire exits to allow quick exit if one is blocked for any reason.

FIRE EXTINGUISHER

Effective firefighting equipment must be available for immediate use. Fire extinguishers of the proper type must be conspicuously located and periodically inspected. Extinguishers containing carbon tetrachloride and other toxic liquids are prohibited. Extinguishers should be recharged regularly following the manufacturer's instructions.

Read labels and follow directions when using fire extinguishers. A fire extinguisher rated not less than 2A shall be provided for each 3,000 square feet of building area, and the maximum allowable distance from any protected area of the building to an extinguisher is 100 feet.

Workers should be trained to use the proper type of extinguisher on fires.

Burning material	Proper extinguishing technique
Wood, paper	Water or any fire extinguisher
Flammable liquid	Smother with dry chemical, foam, carbon dioxide, sand, vaporizing liquid, or other approved fire extinguisher
Electrical fires	Dry chemiccals, carbon dioxide, vaporizing liquid, or compressed gas

EXCAVATING AND TRENCHING

Contact local utilities before trenching or excavating to learn whether electrical or telephone cables or piping are present.

The primary hazard of excavation is a wall collapse. A competent person must 1) assess the "angle of repose" and soil type before digging, and 2) inspect the excavation daily for signs of possible slides or cave-ins. Water and rain increase the hazards of slides and collapses. Excavated material must be stored more than two feet from the edge of a trench. Trenches more than 5 feet deep must be protected from cave-in by shoring, proper sloping of the ground, or another method that meets OSHA regulations. In trenches 4 feet and deeper, sufficient ladders or steps must be provided so no worker must travel more than 25 feet laterally to exit the trench.

FLAMMABLE LIQUIDS

Flammable liquids pose a fire and explosion danger at a work site. Pour gasoline with an approved gooseneck filler. Keep the spout in contact with the metal opening to the tank to prevent sparks from static electricity. Do not smoke while filling. Never fill a machine while it is running. Store gasoline in approved cans away from any possible fire, sparks, or mechanical hazard.

Do not store more than 25 gallons of flammable or combustible liquids inside a building except in an approved storage cabinet. Not more than 60 gallons of such liquid may be stored in any one cabinet.

Outside storage areas may contain no more than 1,100 gallons of fuel in each area. The storage area must be free of other combustible materials and at least 20 feet from any building. "No smoking" signs must be posted in the area.

HAND TOOLS

Wear safety glasses or goggles when using any striking tools. Check that the handle is securely attached before using a hammer. Never pound one hammer with another. Use the correct size hammer for the job. Throw out a hammer, chisel or set with dents, cracks, chips, or mushrooming. Never drive a chisel with a carpenter's or brick hammer; use a sledge or mash hammer instead. Do not drive hardened masonry nails with a carpenter's or brick hammer; use a sledge or mash hammer. Brick chisels and sets are not for use on metal.

Electric hand tools must have double insulation, proper grounding, or a ground fault circuit interrupter. Dead-man switches, which require constant pressure on the trigger to operate, are needed for electric tools on construction sites.

HOIST

Follow manufacturer's directions and limitations. Do not exceed rated load or maximum operating speed. Post hoist safety rules where operators can read them. Material hoists must comply with ANSI A10.5-1969, Safety Requirements of Material Hoists.

Protect entrances to material hoists with full-width gates or bars. Gates or doors of personnel hoists shall be at least 6 feet 6 inches high and protected with mechanical locks which can only be operated by a person in the car. Overhead protective coverings are required on the top of the hoist cage or platform.

HOUSEKEEPING

Good housekeeping reduces hazards and the need to search for tools or materials, and it also increases efficiency by allowing work to flow smoothly. OSHA requires that job sites be orderly and clear of debris and recognized hazards. Material storage and waste disposal yards should be convenient but not in the way. Traffic lanes should permit efficient movement. Store hazardous materials in proper containers and away from the work area. Clean the work areas daily to prevent rubbish from building up. Do not leave lumber with protruding nails in the work area.

LIFTING AND CARRYING

Proper lifting and carrying procedures can prevent falls, back injuries, and hernias. Before carrying something heavy, make sure the route is clear of obstructions and offers good footing. Get a firm grip on the object, being wary of splinters, nails, and other hazards. Wear gloves if needed and wipe off oil or grease before lifting. Then, stand next to the object and lift it with your knees, keeping your back straight and vertical. Set the object down as you lifted it, with your knees and not your back. Make sure your fingers and toes are clear before setting the object down.

When carrying long objects on the shoulder, lower the front end to give a clear view of the path. Use special care when turning. Ask for help when you need it.

MATERIAL HANDLING

1. Lift heavy loads with your legs, not your back.
2. Never overload wheelbarrows, buggies, or hoists. Inspect equipment before use and do not use it if you find a defect.

3. Store masonry units on a paved surface or planks, not directly on the ground. Do not pile bricks more than 7 feet high without supports. When the pile reaches above 4 feet, set bricks back from the edge one inch per foot of height.

4. Stack cut stone on pallets or planks to prevent it from absorbing moisture. Make sure it is well supported to prevent cracking. Thin stone should be stored on edge.

5. Stack cement and lime stacks carefully. Lay each tier at right angles to the one below. Set sacks back one sack's length for every six layers of height: 10 feet is the maximum height unless a bin or other enclosure is used.

6. Stack materials away from hoists, walkways and doorways. Keep materials 10 feet from the edges of floors.

7. Do not exceed the load-capacity of any floor or scaffold that is supporting piles of material.

PERSONAL PROTECTION

Employees must wear and use OSHA-approved safety equipment. Wear steel toes to prevent foot injury. Do not use lifelines or lanyards for any purpose except employee safeguarding.

HEAD

Protective helmets must be worn in areas where head injuries are possible due to flying or falling objects, impact, electrical shock or burns. Helmets for impact protection must meet the requirements of ANSI Z89.1—1969.

HEARING

OSHA sets standards for maximum noise at job sites. If exposure levels surpass these levels, engineering and administrative efforts must be taken to reduce the noise. If these efforts fail, workers must wear hearing protection. Plain cotton is not acceptable for such protection.

Duration per day in hours	Maximum sound level (dBA slow response)
8	90
6	92
4	95
3	97
2	100
1-1/2	102
1	105
1/2	110
1/4 or less	115

Exposure to impact noise must not exceed 140-decibel peak sound pressure level.

EYE AND FACE

Eye and face protection is required whenever there is a danger of injury. This equipment must meet the requirements of ANSI Z87.1—1968 "Practice for Occupational and Educational Eye and Face Protection." Employees working with lasers and welders must wear appropriate filters and safety goggles.

RESPIRATORY PROTECTION

Employers must provide respiratory protection in cases where other efforts to control toxic dusts and vapors are ineffective. Employees must be trained to use this protective equipment. Protective devices must be appropriate to the danger and the nature of the work requirements and inspected daily before use.

VEHICLES

All vehicles must be checked at the beginning of each shift to ensure they are in safe working condition. All defects must be corrected before a vehicle is placed in service. Rollover protection structures are required on most construction equipment, including forklifts and front-end loaders.

No person may use any motor vehicle with an obstructed view of the rear unless:

- *the vehicle has a reverse signal alarm distinguishable from the surrounding noise, or*

- *the vehicle is backed up only when an observer signals that it is safe to do so.*

Heavy machinery, equipment, or other heavy items held aloft must be substantially blocked before employees are permitted to work underneath or nearby.

WALKING AND WORKING ON SURFACES

Obstacles and uneven ground are significant causes of accidents at construction sites. Before a masonry job starts, the area should be cleared and leveled. Fill in low spots and bridge ditches and trenches that remain with planking at least three planks wide. Clear away trash, piles of fill, and other obstacles. Remove ropes, electric cords, and tools from the working area and keep it as clear as possible during construction.

Elevated work areas should have two, widely separated exits. Employees should be aware of safe working practices and allow no deviation from them.

COLOR CODING

Color coding is used to mark areas of particular hazard at some work sites. The technique has long been used as a safety device in factories, with colors painted on machines or walls. OSHA specifies the following color scheme; work sites may use variations as long as the colors are consistent:

Red	Danger—use extra caution. Used to identify emergency exits, extreme hazards, and the location of firefighting equipment.
Yellow—or yellow with black stripes	Exercise caution. Yellow and black is sometimes used to mark traffic lanes.
Green	First-aid supplies, areas of safe refuge, and safety equipment (except firefighting equipment).
Blue	Equipment that must be repaired before use. (Yellow may be substituted in this application.)
Purple	Radiation hazard from x-rays or radioactive materials.

PROTECTIVE RAILINGS

Railings are required on all platforms that are 10 feet or more above adjacent surfaces. Railings are also required on exposed edges of platforms that are narrower than 45 inches and 4 feet or more above the ground. Standard railings have a top rail 42 inches to 48 inches above the floor and an intermediate rail centered between the top rail and the floor. The top rail must be smooth-surfaced. Any point along the rail must be capable of withstanding 200 pounds of pressure in any direction.

A toe board at least 4 inches high is required at floor level to keep tools and materials from falling.

FLOOR OPENINGS

Openings in floors may have standard railings or be covered with planks or steel plate. Do not cover floor openings with plastic as this will hide the danger. Ladderway openings require standard railings except at the entrance. Passage through the entrance must be guarded by a swinging gate or the path can be offset so a worker cannot walk directly into the opening.

RAMPS

Ramps and inclined walkways, used to move workers and materials, are generally constructed of 2×10 or 2×12 lumber. Make sure the wood is well seasoned and free of knots and defects. Fasten the planks with cleats to secure them and reduce springing. A typical ramp angle is 15°. Ramps must be secured top and bottom, and at the middle if needed. Protective railings, 42 inches high, are used on both exposed sides of a ramp where it is more than 4 feet high. Install uprights every 4 feet.

STAIRS

Stairs are preferable to ladders to provide safe access to working platforms. Stairs must be secured at bottom and top. If needed, support the center of the span to prevent collapse. Riser height and tread depth shall be uniform on any flight of stairs. Recommended width is at least 22 inches, tread depth 11 inches, and rise at least 6-1/2 inches. Keep the stairs free of mud, dirt, and water.

Protective railings must be provided on any exposed sides of stairs having at least four risers. Stairways less than 44 inches wide must have railings or other structures on both sides. Stairways between 44 inches and 88 inches wide must have a handrail on enclosed sides and stair rails on open sides.

Stair rails are constructed like standard railings, but the top rail must be from 30 inches to 34 inches from the nose of the tread.

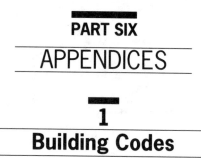

PART SIX

APPENDICES

1

Building Codes

The design and construction of masonry structures are governed by state and local building codes. States generally adopt or "reference" one of three codes: the Uniform Building Code (UBC), the BOCA (Building Officials and Code Administrators International) National Building Code, or the Southern Building Code. States usually institute their own changes in the codes they adopt. In some states, municipalities may adopt codes stricter than the state code.

Building codes can take two approaches:

1. Specification codes describe exactly how a structure will accomplish a certain goal. A typical specification would describe the size and composition of a wall which must bear a certain weight. Specification codes are rigid and must be amended when new techniques or materials are introduced.

2. Performance codes set standards for what the structure must accomplish. Performance codes are becoming more common because they are adaptable to new building materials and methods.

Codes are divided into commercial and residential sections. Residential codes may cover one and two-family housing. Commercial standards, which cover commercial and multifamily structures, are usually stricter than residential standards.

Certain innovations, such as prefabrication of thin brick veneer, are not covered by standard codes and may need case-by-case approval.

Building codes cover the following topics regarding masonry:	
Structural standards	Bearing strength, structural strength, lateral and vertical support, percentage of openings in wall panels.
Veneer	Attachments, backing walls, flashing, and weepholes.
Mortar	Permissible types in various building locations.
Corbelling	How far courses can extend from the wall midline.
Movement joints	Spacing and nature.
Fire	Fire ratings for dividers between different uses of the building.
Fireplace	Construction of firebox; clearance to framing; hearth dimensions; chimney height; multiple uses of flues.
Wood stove	Clearance to unprotected surfaces; hearth; chimney thimble location and construction.
Lintels	Size and bearing surface.
Foundation	Minimum depth below grade; reinforcement.
Joists	Anchoring interval to masonry wall.
Chases	Maximum depth and horizontal length; maximum percentage of wall area.
Beams	Minimum bearing, size, and cross-section.

Building inspectors may be required to make a series of inspections during the progress of the construction. After each inspection, the inspector should fill out paperwork documenting approval of the construction to date, or listing steps that must be taken to bring the job up to code. Keep this paperwork to prevent confusion and misunderstandings in the future.

Do not allow work to proceed beyond the point of a required inspection or you risk covering up parts of the building that the inspector must examine. Inspectors have the right to require contractors to tear down whatever is necessary to permit proper inspection.

Table 4.5 (p. 70) lists allowable compressive stresses in unit masonry, assuming no inspection by an architect or engineer. With testing and/or inspection, allowable stresses are considerably higher.

2
Standards for Masonry

ASTM—the American Society for Testing and Materials—establishes standards for performance and testing of various masonry materials. Standards relevant to masonry include:

ASTM C-55 Standard Specification for Concrete Building Brick

ASTM C-90 Standard Specification for Hollow Load Bearing Concrete Masonry Units

ASTM C-91 Standard Specification for Masonry Cement

ASTM C-129 Standard Specification for Non-Load bearing Concrete Masonry Units

ASTM C-139 Standard Specification for Concrete Masonry Units for Construction of Catch Basins and Manholes

ASTM C-140 Standard Methods of Sampling and Testing Concrete Masonry Units

ASTM C-144 Standard Specification for Aggregate for Masonry Mortar

ASTM C-145 Standard Specification for Solid Load bearing Concrete Masonry Units

ASTM C-270 Standard Specification for Mortar for Unit Masonry

ASTM C-404 Standard Specification for Aggregates for Masonry Grout

ASTM C-476 Standard Specification for Mortar and Grout for Reinforced Masonry

An additional standard for masonry is given by the American National Standards Institute: ANSI A41.1-1953 (R 1970) American Standard Code Requirements for Masonry.

3
Organizations and Associations

STANDARDS ORGANIZATIONS

American National Standards Institute, Inc. (ANSI)
1430 Broadway
New York, N.Y. 10018   (212) 354-3300

American Society for Testing and Materials (ASTM)
1916 Race St.
Philadelphia, PA 19103-1187   (215) 299-5400

Canadian Standards Association
178 Rexdale Boulevard
Rexdale, Ontario, Canada M9W 1R3

National Fire Protection Association
11 Batterymarch Park
Quincy, MA 02169   (617) 770-3000

National Safety Council
444 Michigan Av.
Chicago, IL 60611   (312) 527-4800

Occupational Safety and Health Administration (OSHA)
U.S. Department of Labor
200 Constitution Ave. NW   *(see Table 12.1*
Washington, D.C. 20210   *for phone numbers)*

BUILDING CODE ORGANIZATIONS

Uniform Building Code
International Conference of Building Officials
5360 S. Workman Mill Rd.
Whittier, CA 90601   (213) 699-0541

National Building Code
Building Officials and Code Administrators International
4051 West Flossmoor Rd.
Country Club Hills, IL 60477   (312) 799-2300

Southern Building Code
Southern Building Code Congress International, Inc.
900 Montclair Rd.
Birmingham, Alabama 35213   (205) 591-1853

TRADE ASSOCIATIONS

American Insurance Association
85 John St.
New York, N.Y. 10038   (212) 669-0400

Brick Institute of America
11490 Commerce Park Dr.
Reston, VA 22091   (703) 620-0010

Indiana Limestone Institute of America
Stone City Bank Bldg. # 400
Bedford, IN 47421   (812) 275-4426

International Masonry Institute
815 15 St. NW
Washington, D.C. 20005   (202) 783-3908

Masonry Contractor's Association of America
17W601 14th St.
Oakbrook Terrace, IL 60181   (312) 620-6767

National Concrete Masonry Association
Box 781
Herndon, VA 22070-0781   (703) 435-4900

Portland Cement Association
5420 Old Orchard Rd.
Skokie, IL 60077   (312) 966-6200

Scaffolding, Shoring, and Forming Institute
c/o Thomas Associates
1230 Keith Blvd.
Cleveland, OH 44115   (216) 241-7333

UNIONS

Aluminum, Brick and Glass Worker's International Union
3362 Trollenberg Dr.
Bridgeton, MO 60344   (314) 734-6142

International Masonry Institute of Apprenticeship and Training
823 15 St. NW Suite 1001
Washington, D.C. 20005   (202) 783-3908

International Union of Bricklayers and Allied Craftsmen
823 15 St. NW Suite 1001
Washington, D.C. 20005   (202) 783-3788

4
Glossary

Absorption. The amount of water a masonry unit can absorb. The percentage increase in weight measured when a unit is submerged in cold or boiling water for a set time.

Abutment. Section of wall adjacent to an arch which sustains the lateral and vertical thrusts of the arch.

Accelerator. A chemical additive to make cement or mortar set faster.

Adhesion. Ability of mortar to stick to various materials, usually a masonry unit.

Admixture. A material added to mortar or concrete to change its character, such as its color or setting rate.

Aggregate. Material, usually sand or gravel, which is combined in mortar or concrete to form the structure.

Anchor. A metal device used to fasten two building elements together. Used to bond facing wythes to backing walls or to fasten other elements to masonry.

Angle bolt. A bolt set into mortar, grout, or concrete for attaching another building element to masonry or concrete.

Angle of repose. Slope of soil surrounding a trench or excavation.

ANSI. American National Standards Institute, an organization that sets standards for materials.

Antifreeze. Chemical added to mortar or concrete to prevent freezing.

Arch. A (usually) curved structural element used to span an opening without using a lintel.

Ashlar. Stonework composed of squared stone with tops and bottoms laid horizontally.

ASTM. American Society for Testing and Materials. An organization that writes procedures for testing materials.

Backfill. Material used to fill around a foundation.

Backhand. Method of laying brick while moving backward.

Backing up. Laying the backup wall after the facing wall.

Backparging, backplastering. Applying a thin coat of mortar to the back of one wythe in a composite wall.

Backsight. A sighting with the builders' level or transit on a point of known elevation.

Backup. Wythe behind a face wythe used for structural support, additional weatherproofing, and insulation.

Base flashing. Flashing inserted into shingles and bent up the side of a wall or chimney.

Base plate. A device that distributes the load of a scaffold to the ground.

Bat. A part of a brick.

Batter board. A pair of horizontal boards at right angles to each other that are placed outside the corners of a foundation during layout. Used to mark building corners and lines.

Batter. To slope the outer side of a masonry wall.

Beam. A horizontal structural member.

Bearing. A surface that supports a load.

Bearing plate. A steel plate that distributes the load of a support member, such as a beam.

Bearing wall. A wall that supports part of the building weight (not just its own weight).

Bed. The bottom side of a brick or block as it sits in the wall.

Bed joint. The horizontal mortar between courses.

Belt course. A single, usually decorative course, of contrasting or corbelled units.

Bentonite. A fine clay used to waterproof below-grade structures.

Bond. 1) The pattern of bricks or blocks in a structure. 2) The adherence of mortar to masonry units. 3) Joining of two wythes.

Bond beam. A continuous beam of reinforcing and grout used to tie masonry structures horizontally.

Bond stone. A unit that spans two wythes to join them structurally.

Breaking the joint. Covering a head joint with another masonry unit.

Breast. The portion of a fireplace front above the opening.

Brick ledge. The protrusion on a foundation to hold a brick veneer.

Butter. To apply mortar to the head joints of a unit before laying.

Buttress. Masonry projection to support and stabilize a wall.

Cavity wall. A wall in which the two wythes are separated by a small air space. Also "hollow wall."

Cell. Opening in a masonry unit with area greater than 1-1/2 square inches.

Cementitious materials. Portland cement and lime portion of mortar.

Center or **centering.** Form used to lay an arch.

Ceramic glaze. A glass-like coating fused to the surface of tile or brick. Available colored or clear.

Ceramic tile. A thin tile set with adhesive or mortar. Usually set by a tile setter, not a mason.

Ceramic veneer. Terra cotta tiles set by anchors and grout, or by adhesive.

Chase. Recess provided for other building elements, such as pipes or ducts.

Class A chimney. A residential chimney with a flue used for any solid fuel, gas, oil, or propane.

Clinker brick. A distorted or discolored brick produced by excessive heat during burning.

Closure (also closer, closure brick, closure block). The last unit laid in a course.

CMU. Concrete masonry unit.

Collar joint. Gap between backing wythe and face wythe.

Composite wall. Wall constructed with different material in each wythe.

Compression. Downward loading on a building element.

Compressive strength. The ability to withstand compression.

Control joint. A joint that can slip to allow for thermal expansion and shrinkage drying in concrete masonry.

Convection. A physical force by which denser fluids fall and less dense fluids rise. Causes the rise of smoke in a chimney.

Coping. Waterproof cap for a wall or pier.

Corbel. Courses that extend out from the wall plane. Used as a decoration in chimneys or to support flue in a fireplace smoke chamber.

Core. The empty center sections of a block or brick.

Counterflashing. Flashing inserted into joints in chimneys and walls to cover the base flashing.

Creeper. A brick sawed or cut to fit above the arch and beneath the full courses.

Crown. Highest point of an arch; also curve in an exterior paved surface for drainage.

Cull. A substandard masonry unit; also the top surface of a brick.

Curtain wall. A nonbearing exterior wall that is supported at each end but does not carry structural loads.

Cutting plane. A line in a blueprint indicating where a section drawing was taken.

Damp course. A layer of impervious material to prevent moisture from entering a structure.

Denominator. The bottom number in a fraction; the number that names the type of fraction. The "4" in one-quarter.

Dentil. A projection left by using square units in a nonsquare corner.

Dimensioned stone. Stone cut to a specific size.

Draft. The force of convection which pushes smoke up a chimney.

Drip. Projection on a unit to prevent water from dripping down the wall. A cut in the underside of such a projection.

Dry. Built without mortar, as in dry wall, dry joint, and dry bonding.

Efflorescence. Powder or stain on masonry surface caused by salts left behind when water evaporates.

Expansion joint. A joint used to allow expansion in brick masonry.

Extrados. The upper curve of a curved arch. See intrados.

Face shell. The outer wall of a concrete block, as distinguished from the webs that join the shells together.

Face. The exposed, visible surface of a wall or unit; the visible surface of a fireplace.

Face shell bedding. Laying blocks with mortar on the face shell but not on the webs.

Factor of safety (safety factor). The ratio of load at failure to working load. Used to evaluate scaffolds and hoists.

Fat mortar. Mortar with a high proportion of cementitious materials. Sticks to the trowel better than lean mortar.

Fire wall. A wall designed to resist the spread of fire.

Firebox. The burning chamber of a fireplace.

Firebrick. Bricks made to resist high temperatures for use in fireboxes and kilns. Also "retractory" brick.

Fireclay. See refractory cement.

Flagstone. A flat stone often used in walkways. Each stone is called a "flag."

Flash set. Extremely rapid set or hardening of mortar or concrete.

Flashing. Flexible material used to prevent water from entering a structure or to divert water that has entered to the outside.

Forehand. Laying brick while moving forward along the wall. See backhand, overhand.

Foresight. A sighting with builders' level or transit on a point of unknown elevation.

Frog. Small depression in the bed of a brick.

Frost line. Depth at which earth can freeze in the coldest years in a building location.

Full shell bedding. Laying blocks with mortar on face shells and webs.

Furring. Strips of wood fastened to masonry to support lath, insulation, wallboard, or paneling.

Furrow. To groove the surface of freshly laid bed mortar before placing bricks. Prohibited by some building codes.

Green. Mortar or concrete that has set but not cured fully.

Grounds. Strips of wood set into masonry walls for future attachment of trim or furring.

Grout. Soupy concrete or mortar used to bond reinforcement and make bond-beams. Also mortar with fine aggregate used to fill gaps in tile.

Hard to the line. Units set too close to the mason's line.

Harsh mortar. Mortar that is difficult to work.

Hawk. A square mortar board mounted atop a handle used for plastering and tuckpointing.

Head. The top of a window or door frame.

Head jamb. The top element of a door or window frame.

Head joint. A vertical mortar joint between units.

Header. Unit laid on its bed with the end facing out. Often used to bond two wythes together.

Heel. The edge of a retaining wall footing on the uphill side.

High lift grouting. Technique of pouring grouting into a full story of wall in one series of operations.

Hub. A stake marking the corner of a building during layout.

Hydrated lime (slaked lime). Quicklime that has been "slaked" with water; used to increase workability of mortar.

Hydrostatic pressure. Pressure created by the weight of water above. May force water to leak into below-grade structures.

Hypotenuse. The side of a right triangle opposite the right angle.

Initial rate of absorption (IRA). Grams of water absorbed by 30 square inches of a submerged brick in one minute.

Intrados. Lower or inner surface of an arch. See extrados.

IRA. See Initial rate of absorption.

Jamb (side jamb). The column of units adjacent to an opening; the side of a window or door frame.

Key. A rectangular-section protrusion below a retaining wall footing to prevent downhill slippage.

Keystone. The center brick or stone in an arch.

King Closure. A brick with a corner clipped off.

Laitance. Surface layer of fines on concrete or mortar brought up by excess water in the mix.

Lap. Distance by which one unit overlaps the one below it.

Lateral force. A force that pushes a structural element in a direction parallel to the ground.

Lead. A series of courses laid before the main structure to serve as a guide.

Leveling rod. A rod with a scale used with a transit or builders' level for shooting heights.

Lift. The amount of grout poured in a single operation. Several lifts can make one pour.

Lime putty. Hydrated lime that is delivered wet to the job site. Was used for making mortar before lime was available in bags.

Line of flight. A line joining the tread noses of a flight of steps.

Lintel. A horizontal beam used to span an opening.

Low lift grouting. Grouting a wall after small portions have been built.

Masonry unit. A building unit of concrete, ceramic, glass, gypsum, or stone.

Middle lead. A support at the middle of along course to prevent sagging of the mason's line.

Modular dimension. The size of a unit after it is laid in mortar; equal to unit dimension plus one joint.

Mud sill. A footing that distributes the load of a scaffold to the ground. Usually made of wood.

Muriatic acid. Mason's name for hydrochloric acid, used for cleaning masonry.

Non-bearing wall. A wall that supports only its own weight.

Numerator. The top number in a fraction; identifies the number of units of the denominator in that fraction. The "1" in one-quarter.

OSHA. Occupational Safety and Health Administration. Federal agency charged with overseeing safety and health at workplaces.

Outer hearth. The section of masonry extending out from a fireplace face. An extension of the inner hearth.

Overhand work. Laying brick from inside the wall.

Panel. A section of wall between supports, such as piers or pilasters.

Parapet. A wall that extends above a roof.

Parging. Applying a coat of mortar to a masonry wall, generally to increase water resistance. Also the material used for this process.

Partition. An interior wall, usually nonbearing.

Paver. A masonry unit made for paving.

Pencil rod. One-quarter inch diameter reinforcing rod.

Pier. A masonry or concrete column to support a floor structure in buildings with no basement; a column used to support garden walls; a freestanding, rectangular column.

Pigeonhole. A void left by using square units to form a nonsquare corner in a wall.

Pilaster. A pier or column projecting from a wall, designed to receive a load of a beam or to strengthen the wall against lateral forces.

Plumb. Precisely vertical.

Pointing. Process of filling voids after units have been laid.

Pour. The amount of grout poured in a short period, before the grout sets. May consist of several lifts.

Puddling. Settling of grout or concrete by moving the mixture about with a rod or shovel.

Putlog. A separate horizontal member that carries a load in a scaffold and bears on the wall being scaffolded.

Pythagorean theorem. A formula describing the relationship between the three sides of a right triangle. $A^2 + B^2 = C^2$.

Quarter closure. A brick cut to one-quarter its full length.

Queen closure. A brick next to a corner in some pattern bonds; equal to one-quarter brick in length.

Quicklime (lime, hot lime). Caustic lime made by heating limestone. When water is added, lime putty results.

Quoin (pronounced "coin"). A large stone, brick, or block that forms a corner. Often projects from the wall surface.

RBM. Reinforced brick masonry. Brick masonry incorporating steel reinforcing.

R-value. A measure of resistance to flow of heat. Used to describe the effectiveness of insulation. Higher values indicate greater insulating power.

Reglet. A groove or slot in a unit. Used to receive the edge of a flashing.

Rake back (rack back). To end each successive course one-half unit further back from the course before.

Rake-back (rack-back) lead. A section of wall laid in the middle, at the same time as the corner leads. Used to stabilize the mason's line while the rest of the wall is being laid.

Rake. To clean mortar from a joint to an even depth.

Reference point. A location established by an architect or engineer as the starting point for the layout, both in altitude and position.

Refractory brick. See Fire brick.

Refractory cement. Special heat-resistant mortar used for setting firebricks, flues, and other high-temperature material. Also "fire clay."

Retarder. Admixture used to slow the set of mortar or concrete.

Retemper. To add water and remix mortar to make it more workable.

Return. A surface at (usually) 90° from the principal face of a unit or structure, at a jamb, for instance.

Rip block. A block that is less than full height.

Rise. Difference in height.

Riser. The vertical section of a step.

Rolling tower. A scaffold system riding on casters.

Rowlock (rolok). Unit laid on face with the end facing out.

Rowlock stretcher. Unit laid on its side with the bed facing out.

Rubble. Masonry built with rough stones; also masonry debris.

Run. Horizontal difference in location between two points.

Sailor. Brick laid on end with the bed facing out.

Salmon. An underfired brick, now only available as salvage brick. Has low compressive strength and poor resistance to weather.

Scratch coat. A first coat of a multicoat plastering or parging operation. When partly set, the coat is scratched so the second coat will adhere better.

Serpentine. Wall that curves back and forth.

Set. Hardening of mortar. Initial set gives mortar some strength; final set gives it full strength.

Shoot heights. The process of figuring out the relative height of various spots on a building site.

Shop drawing. Detail drawing made by contractor or manufacturer.

Sill course. The course of bricks which support a sill, or rowlock, course.

Skewback. Surface where an arch meets the supporting wall.

Slack to the line. Units laid too far from the line.

Slushed joint. Process of adding mortar to a vertical joint after units are laid. An unprofessional practice leading to poor bonding and leaky walls.

Smoke test. Building a fire in a fireplace with the chimney top closed to check for leaks.

Soffit. Underside of a building element, such as an arch or eave.

Solar storage. Storage of sun's heat in massive building element, such as a masonry wall or floor.

Soldier. A brick or block set on end with the face toward the outside.

Spall. To deteriorate on the surface.

Span. Distance between supports; width of a bay in a serpentine wall.

Spandrel. Triangular area on the sides of an arch; an edge beam between columns or piers along the outside of a building.

Splay. A surface with a slant or a bevel.

Split. A brick cut to half the height of a full unit.

Split lintel. A precast concrete lintel designed to be used in pairs in multiwythe walls.

Story pole. A pole marked with heights of courses and openings; used to lay courses accurately.

Stretcher. A unit laid with its face out and bed down. The normal position for units.

Substrate. Backing for plaster. Can be lath or masonry.

Suction rate. See Initial Rate of Absorption.

Surface bonding cement. Material containing fibers used to lay blocks without mortar trowelled on the surface.

Tail. One of the two lines of units comprising a corner lead.

Template. Pattern cut to simplify laying complicated shapes, such as curves.

Tensile strength. The ability to withstand tension.

Tension. Force that tends to pull a material apart.

Terra cotta. Fired clay units, usually with glazed surface. May be ornamental or structural.

Three-quarter closure. A brick cut to three-quarters length.

Tie. Masonry unit or metal device used to secure two wythes together or to fasten a masonry wall to a backing. Also a device to link a scaffold to a building.

Toe. The portion of a retaining wall footing on the downhill side.

Toeboard. A barrier at the side and end of a platform to prevent material from falling.

Tooling. Treating a joint after the units have been laid to improve appearance and weather resistance.

Toothing. Temporary end to a wall. Units project in a zigzag fashion so the completed wall can be bonded later.

Total rise. The vertical measurement of a set of steps.

Total run. The horizontal measurement of a set of steps.

Tread. The horizontal section of a step.

Trig (twig). A support that holds a long mason's line so it does not sag or deviate in the center.

Tuckpoint. To repair deteriorated joints by cleaning them and forcing in new mortar.

Veneer. A wythe of masonry attached to a backing wall to make a facing. Usually has no structural value.

Vitrified. Glassy appearance. Seen in terra cotta units and sewer pipe.

Wall panel. Wall section laid on the ground or in the shop and hoisted into position.

Wall plate. The wall of a retaining wall, exclusive of pilasters supporting it.

Wall tie. Metal tie used to join two wythes or one wythe to a nonmasonry backing.

Water retentivity. Ability of mortar to retain water and remain workable.

Web. Sections of concrete block that join the face shells.

Weephole. Void at the bottom of a wall to allow water to exit the wall or structure.

Working drawing. Blueprint.

Wythe. An individual tier of masonry units. A veneer wall contains one wythe, a solid masonry wall often contains two or more.

INDEX

Anchoring bearing plate, 268
Acid cleaning, 189-192
Anchors, 33-38, 154-157
   Adhesive, installing, 156-157
   Driven, installing, 157
   Expanding, 155-156
   Selection, 36
Angle bolts, installing, 155
Angle iron, 81
Angles, measuring, 318
Apprenticeship program, 158
Arches, 158-166
   Building, 164-166
   Centers, 161-166
   Layout, 160-164
   Terminology, 159-160
Area measurement, 318-321
Associations, 341-343

Barbecue, 166-167
Batter boards, 282-284
Bearing plate, installing, 268
Bearing, soil, 233-234
Bentonite, 56
Bituminous sealer, 57
Block
   Chimney, 44
   Concrete, 38-44
   Glass, 44-45
Blueprints, 167-174
   Abbreviations, 173-174
   Lines, 168-169
   Symbols, 171-172
   Views, 169-171
Bond beam,
   Intersections, 110
   Properties, 95
   Types, 96
Bond break, 175
Bonds, 175-184
   Brick, 175-179

   Concrete block corners, 211
   Pattern, 179-184
      Basketweave, 183-184
      Herringbone, 181-183
      Square, 179-181
Brick, 46-54, 132, 209
   Absorbency, 47
   Bonds, 175-179
   Concrete, 175-179, 209
   Firebrick, 132
   Grade, 48
   Size, 48, 51
   Types, 49-54
   Terminology, 49-50
Brushes, 1-3
Building codes, 338-339, 341-342

Caulking
   Application, 184-186
   Coverage, 55
   Types, 54-55
Chalk line, 3
Chases, 186-187
Chimney, 141-153
   Categories, 142
   Creosote and chimney fire,
      151-152
   Downdraft, 152-153
   Flashing, 145-147
   Laying, 149-151
   Parts, 143-149
   Reinforcing and anchoring,
      142-143
   Smoking, 153
   Troubleshooting, 151-153
   Water resistance, 143
Chisel, 3-4
Cleaning, 187-195, 215-216
   Acid, 189-192
   After laying, 215-216
   Dry, 193-194

357

New masonry, 190-191
Wet, 189-193
Coatings,
Application, 195-204
Cementitious waterproofings, 197
Clear coatings, 203
Membrane waterproofing, 203-204
Parging, 198
Plaster and stucco, 198-200
Preparation, 196-197
Choosing, 58-59
Types, 56-60
Coloring mortar, 204
Columns, 205-206
Concrete block construction, 206-211
Bond beam, 209-211
Laying units, 207-209
Concrete, 60-61
Construction practices, 211-216
Control joint, 67-68
Coping, 61, 216
Corners, 217-222
Convex and concave, 221-222
Leads, 217-219
Obtuse and acute, 218, 220
Courses, spacing, 222-224
Cracking, 262, 294
Curing, 215-216
Cutter, 4
Cutting plane, 171
Cutting material, 224, 228

Diagonals, 279
Doors, 228-230
Drain tile, installation, 234-235
Drill, 22

Efflorescence,
Causes, 294
Removing, 194-195
Elastic compounds, 57
Electric cord, 22

Estimating, 309-317
Block, 314-316
Brick, 310-312, 315
Mortar, 312-314
Ties and reinforcement, 316-317
Expansion compounds, 57
Expansion joint, 68

Fans, 308
Firebrick, laying, 132
Fireplace, 128-140
Dimensions, 129
Draft, 131
Equipment checklist, 138-139
Outdoor, 166-167
Parts, 132-138
Planning, 139
Rumford, 128-129
Safe operation, 140
Flashing,
Application, 231-232
Chimney, 145-147
Material, 62
Window, 308
Float, 4-5
Flue, 63-64, 147-148
Footings, 233-236
Forklift, 23

Glass block
Material, 44-46
Installation, 236-239
Grinder, 23
Grouting, 240-244
High lift, 243-244
Low lift, 242
Preparing, 241

Hammer, 5
Hoist, 24
Hydraulic cement, 57

I-beam, 81
Insulation
Exterior systems, 65-66

Pourable, 64
Rigid, 65

Jointer, 5-6
Jointing, 248-251

Ladder, 29
Landscaping, 252
Laying,
  Units, 214-215
  To the line, 252-253
Layout tools, 6-11
Layout, 275-284
Leads, building, 217-219
Level
  Automatic, 8
  Builder's, 7, 276-284
  Levelling rod, 9
  Line, 10
  Spirit, 10
  Transit, 7, 276-284
  Water, 11
Lime, 73
Line and line stretchers, 11
Lintels, 253-258, 307
  Brick, 254
  Concrete, 255-257
  Sizes and types, 81-83
  Steel, 257-258

Masonry cement, 73-75, (see mortar)
Masonry guide, 13
Material, storing, 214
Mathematics, 318-321
Measurements, 322-327
  Engineer's measurement, 321-324
  Metric conversions, 323, 325
Miscellaneous tools, 12
Mixing equipment-hand, 14
Modular planning, 258
Moisture
  Resistance, 195-204
  Leaking crack, filling, 200-201

Problems, 297-300
Mortar board and tub, 14
Mortar mixer, 24
Mortar, 68-78, 258-261
  Admixtures, 71-72, 300-301
  Batching, 259
  Checking the mix, 260-261
  Cold and hot weather, 297, 300-304
  Coloring, 72-73, 204-205
  Mixing, 259-261
  Properties, 69-71
  Retempering, 261
  Special purpose, 77-78
Movement
  Differential, 262
  Expansion and contraction, 294-295
  Settling, 295
Movement Joint, 244-248
  Control joint, 246
  Expansion joint, 246-247
  Locating, 245-246
  Sealing, 247-248

Paint and coatings, 59-60, 201-204
  Latex, 201-202
  Portland cement, 201
  Surface preparation, 202
Parging, application, 198
Paving
  Material, 79-80
  Mortared and mortarless, 262-265
Piers, 266-268
Pilaster, 268-271
Plaster, 198-200
Plumb bob, 15
Portland cement, 80-81
  Paint, 201
Power washer, 24

Reinforcing bar (re-rod), 83-84
Reinforcing, 81-85
  Placement and use, 271-272
Repairs, 272-274

Rulers, 15-16

Safety
  Equipment, 16-17
  Scaffold, 27-32
  OSHA offices, 328
Sand, 75-76
Sandblasting, 24, 194
Saw, 25
Sawing, 224, 228
Scaffolding, 27-32
Schedules, 175
Shooting angles, 277-278
Shooting heights, 276-277
Silicon sealer, 58
Site preparation, 274-275 (see layout)
Spacing rule, 222-224
Specifications, 172
Square, 17
Stains, 194-195
Standards for masonry, 340-341
Steel plate, 84
Steps, 284-290
  Laying, 288-290
  Planning and layout, 285-287
  Terminology, 284-285
Stone masonry, 85-87, 290-293
  Artificial, 292-293
  Lifting slabs, 293
  Veneer wall, 290-291
Stone tools, 17-18
Story pole, 18
Strap anchor, 84
Stucco, application, 198-200
Stud tool, 26

Terra cotta, 87-88
Tile, 88-89
Tongs, 18
Tooling, 215-216, 251
Transit, 7,8
  Use, 276-281
Trigs, use, 252-253
Troubleshooting, 151-153, 293-300
  Chimney, 151-153

Cracking, 294
Efflorescence, 294
Expansion and contraction, 294-295
Settling, **295**
Spalling, 295-296
Stains, **295**
Water, 296-299
Trowel, 18-21
Tuckpointing
  Mortar, 78
  Technique, 274

Veneer
  Stone, 290-291
  Brick, 125-127
Vents, 308

Wall area, 310
Wall ties, 90-92
Walls, 94-127
  Backfilling, 120
  Butressed, 118-120
  Cantilever, 117-118
  Cavity, 94, 96-99
  Checklist, 100
  Composite, 101-102
  Counterfort, 118-120
  Foundation and basement, 102-104
  Garden, 105-107
  Gravity, 116-117
  Intersections, 107-111
  Parapet, 111-112
  Retaining, 112-120
    Water control, 115-116
  Screen, 120
  Serpentine, 121-125
  Veneer, 125-127
    Thin, 127
Water, mixing, 76
Water, problems, 294-300
Weather, 297, 300-304
Wheelbarrow and carts, 26
Windows, 304-307